ENGINEERING AN AIRLINE

Paul Jarvis

AMBERLEY

In association with **BRITISH AIRWAYS**

First published 2017

Amberley Publishing
The Hill, Stroud
Gloucestershire, GL5 4EP

www.amberley-books.com

ISBN 978 1 4456 6706 5 (paperback)
ISBN 978 1 4456 6707 2 (ebook)

British Library Cataloguing in Publication Data.
A catalogue record for this book is available from the British Library.

Typesetting by Amberley Publishing.
Printed in the UK.

CONTENTS

ACKNOWLEDGEMENTS

I've known many engineers in my fifty-year career in British Airways. My father who was a licensed ground engineer for BOAC took me first to their engineering Christmas parties under the wing of an aircraft in a hangar at Heathrow, a magical experience for an eight-year-old, and I've long suspected that 'Santa' was one of his engineering colleagues. I joined BOAC engineering myself at twenty-one, very low down in their hierarchy, but gradually rose by a circuitous route to be their company secretary. From clerks and secretaries to chartered engineers and directors, it's been an incredibly interesting experience, for me at least, and the story of airline engineering I've always felt needed telling.

What better chance now as curator of the British Airways Heritage Collection to set out the significant contribution that engineering has made to the story of British Airways and its predecessors. What became clear to me early on in my research, however, is that this is a story that is as deep as it is wide and not so easily distilled into a relatively modest-sized book. As I'm also not an engineer, I've had to take advice and guidance from current and former engineering colleagues and I hope they will accept some licence from me in summarising the often detailed technicalities of what aircraft engineering entails, even from the earliest days of wooden biplanes.

This book, then, represents an accumulation of many years of experience and knowledge, not least from my team of volunteers at the British Airways Heritage Collection several of which are, fortuitously, former engineering colleagues. My special thanks go to Jim Davies, Alan Cavender and Keith Haywood for advice and diligent proofreading and to Christine Quick and Adrian Constable for scanning and improving the many photographs.

PJ
20 March 2017

FROM SHEDS TO A LISTED BUILDING

When I joined BOAC, one of British Airways' predecessor airlines, in 1966, it was as a technical clerk in the Engineering Department. I was very low down in the pecking order of things but I was given a very wide freedom to carry out my job dealing with the many engineers I met on a daily basis when amending the many technical manuals in the myriad of hangars and workshops that comprised BOAC's engineering base at Hatton Cross on Heathrow airport. I even thought of trying to progress further in engineering. My father was a licensed ground engineer for BOAC so it was in the blood so to speak. Despite being offered a job as a technical illustrator, I decided a role in the commercial side of the airline would offer more prospects and, after a year, moved on. That one year, however, had a lasting impression on me. In part it was the fascination of being, on a day-to-day basis, very close to large commercial aircraft and, in my own very small way, being part of a highly experienced team of (almost entirely) men inspecting, maintaining and mending very complex and intricate pieces of flying machinery.

Now here's a surprising thing. Apart from articles now and again in in-house staff newspapers and magazines, one rarely read about ground engineers in those early days despite their central and vital role in keeping the airline flying and on time. I'd certainly not seen a book devoted to their work. Not that they were a forgotten breed, but working, often unseen or unrecognised, often in less than clean overalls and wheeling a trolley of spanners in their daily grind, they did not carry quite the same glamour that other airline staff, especially uniformed members, experienced and enjoyed. Engineering as a profession was also, quite wrongly, often looked down on in mid-20th century society; it was something that those 'good with their hands' did and that by implication they were somehow lesser members of society. Fortunately, and well before the turn of the century, engineers in all walks of life have come out of the shadows and it is a sought-after profession, especially in aviation, that many now enter regardless of gender.

This book is an attempt to bring into the daylight some of the history of British Airways Engineering, including its predecessor airlines, going back to the 1920s and the days of Imperial Airways and beyond. It is not a book for engineers but the general reader so I hope I'll be forgiven for skimming over a lot of detail. There are, for example, no detailed explanations of how to maintain a flying boat (there actually is a book out there that explains all of that), but it does set out in a more broad brush sort of way what is a very interesting story.

Taken from the beginnings of commercial civil aviation, we might wonder today at some of the practices and procedures applied in yesteryear, but they are part of the foundation of what is today a cornerstone of the commercial airline world and a very complex and highly skilled engineering business. I also use the term 'engineer' in its broadest sense. Throughout British Airways' history there have been many sub-sets of 'engineer' with that title often reserved for those in the highest grades and with the highest qualifications, such as 'licensed aircraft engineers'. Others might have job descriptions with titles as varied as mechanics, tradesmen, technicians and maintenance workers among others. There were also until the 1960s flight engineers, often a ground engineer retrained to support the pilots in managing an aircraft's complex systems, 'flying spanners' as they were affectionately known. Nowadays, in 21st-century collective parlance, all are engineering 'colleagues'. For simplicity, I prefer to use the term 'engineer'.

There's an interesting set of photographs in the British Airways Heritage Collection dating from around 1922. They show that while aviation had advanced considerably during the First World War it was still a fledgling industry. The engineers of the time used facilities not much different from the corrugated iron and wooden garages my father and his father before him had often serviced their cars in during the 1920s and 1930s, and often using the same tools. An engine was an engine and could be lifted out of a car using the same ropes or chains used in the hangar. The aircraft engine was just a lot bigger than the one in my father's Ford Anglia.

In 1922 civil commercial flying had only been operating for just over three years and it was a time of intense competition between many small private airlines chasing not many passengers. Costs, not least engineering costs, were high. While there remained a huge surplus of relatively cheap ex-military aeroplanes and spare parts, the rate of aircraft depreciation was measured in a few years at best. Aircraft life spans were short given their construction was flimsy and the technology

of the time was basic by today's standards. Military aircraft were also not expected to last long, a few weeks at best. Aircraft engines were also notoriously unreliable and needed constant maintenance. A good mechanic, as they were called in those days, was a valuable employee and often flew as a second crew member to carry out any necessary repairs en-route, the original 'flying spanners'. Then, as now, operating an air service to a timetabled schedule was very important. This may have been an embryonic industry but passengers did not expect to be kept waiting because of the late delivery of a serviceable aircraft from its maintenance checks. Gaining a reputation for unreliability was a pathway to bankruptcy with both UK and foreign competitors ready to step in. The more traditional forms of transport operators such as the railways and shipping companies would also remain strong competitors to aviation for many decades yet to come.

When I joined BOAC fifty years ago there still remained at Heathrow a motley collection of what can only be called 'sheds', in which a variety of small engineering maintenance works were carried out. I well remember the BOAC brake shop and its black and grimy interior (cleaning aircraft brake pads was a notoriously dirty job and very grimy overalls were standard wear). Not much glamour here. BOAC had built these sheds in the late 1940s as part of its new engineering base at the then new Heathrow airport that had opened for business in May 1946. I don't know what the brake shop had originally been built for but twenty years later it was still active and a throwback to the sheds of the 1920s where aviation had first began. In the early 1950s, about a quarter of a mile away from the brake shop, BOAC had also had built a very large new hangar known as 'Headquarters Building'- a rather military way of describing where the vast majority of the airline's staff were to be based, not least its engineers. This was no throwback to the

Previous spread, left: By 1969, when this overhead photograph was taken, the BOAC and BEA engineering bases at Heathrow had grown into a huge and complex multiplicity of hangars and other buildings in only twenty-five years since the airport first opened. This growth was forced by a rapid increase in the size of the aircraft fleets of the two airlines as their businesses developed during the 1950s and 1960s. There seemed never to be quite enough hangar space, especially as technology advanced and aircraft became larger. Although BOAC and BEA were both UK nationalised airlines they were entirely separate businesses operating different route networks. Very little of their individual premises were shared and so there was considerable duplication of workshops and other facilities, a major task for British Airways to sort out when it was formed in 1974 from the merger of BOAC and BEA and had to amalgamate the two companies' businesses. This area is in the south east corner of the airport alongside the A30 London road near the Hatton Cross Piccadilly Line Underground station. On the left of the photograph are BEA's two large hangar complexes, known as West Base in later years, and to the right BOAC's, known as East Base. Technical Block 'A' (TBA), where I started work, is the large building in the top right of the image and was for many years BOAC's principal engineering facility. Elsewhere are scattered a myriad of older hangars and buildings from the 1940s. It was a maze then but less so today with many of the old hangars and buildings now demolished to make way for aircraft parking.

Right: AT & T, a predecessor airline of British Airways, employed its first two licensed engineers in early 1918, Messrs Woodham and Kelly. They did not just maintain aircraft on the ground but also in the air and they were a breed of men key to keeping AT & T's services flying and on time. They must have been a bit of a cross-breed between ground engineers who never flew and those who did from time to time. In fact, they became the first of the early flight engineers whose principal role was to fly with the pilot and keep the aircraft serviceable during its round-trip journey. The early aircraft were unreliable and often needed their engines adjusting en-route, an appropriate job for an engineer. The pilot had to find a convenient field to land in so the engineer could hop out, adjust whatever was needed and take off again; quite an experience for the airline's passengers.

1920s but a state-of-the-art hangar and building complex in the spirit of London's 1951 Festival of Britain, a festival of renewal, technology and optimism that was to herald in the end of 1940s austerity. Headquarters Building reflected all that despite its rather grey and austere external appearance. In 1965, just before I joined, BOAC decided to rename most of its Heathrow buildings and Headquarters Building became known as 'Technical Block 'A', shortened to 'TBA', to which it is referred to this day.

The new building gave BOAC not only a flagship headquarters but the latest facilities in which to engineer its aircraft. Its four interconnected hangars were vast, each fronted by huge unsupported concrete archways considered at the time the widest unsupported concrete spans in the world. In April 1996 the spans were listed as Grade: II by English Heritage, not quite what one might expect with a more traditional building such as a thatched cottage or half-timbered mansion more in mind, but architecturally considered worth preserving. The old sheds had been demolished well before then to make way for BOAC's new computer building, 'Boadicea House', and the renamed TBA might also have fallen to the same fate on the grounds of being unsuitable for 21st-century aircraft maintenance. Its listed designation, however, has preserved it and it has taken on a new lease of life in recent years. We'll come to that subsequently.

BEA, British European Airways, also a predecessor airline of today's British Airways, had, by the mid-1950s, also established its new engineering base at Heathrow, a somewhat different style of building complex to TBA, being more contemporary in style and shape. Maybe that's why it was never listed but it fitted BEA's engineering purposes very well. At that time BOAC and BEA were the UK's two nationalised airlines but they operated quite different route structures. BOAC's aircraft could be away on service for up to twenty-one days at a time having to be maintained and serviced at many far-away places, often on the other side of the world; quite a bit of glamour here with overseas engineering postings, especially to places such as the Far East and the Caribbean, coveted and sought after. BEA's aircraft were referred to as 'Back Every Afternoon', a play on the company name BEA, in that they only operated from the UK to European cities or on UK domestic routes. Provided any aircraft technical problems en-route were not considered unsafe, BEA's aircraft could fly home and its unserviceable components fixed overnight back at base at Heathrow.

The two companies' engineering organisations and building complexes were therefore established to reflect their different operations and, inevitably, different working practices and procedures. I'm sure BEA had a brake shop, maybe in a shed somewhere, but I never saw it. I was a BOAC man and we did not venture into their territory without being invited. In effect, the two companies operated quite separately and independently and were sometime strong rivals with similar levels of engineering expertise in many areas and facilities to match.

Opposite: What we would now describe as a major privilege was the ability in the early days of my career just to stroll into the hangars and wander around whatever aircraft was in for maintenance. As long as you checked with the supervisor he would normally allow you on board but often with the warning to 'watch out for the holes in the floor'. It was a major lunchtime activity for my office colleagues and I and we'd often take external guests along. I once left my briefcase in the Concorde hangar on a table when taking some European politicians around (I was in British Airways' government affairs department in the 1980s) only to rather red-facedly scurry back praying the security alarm did not go off. Nowadays, and quite rightly, it is a major security exercise to allow anyone in the hangars or anywhere near an aircraft without being closely escorted and supervised.

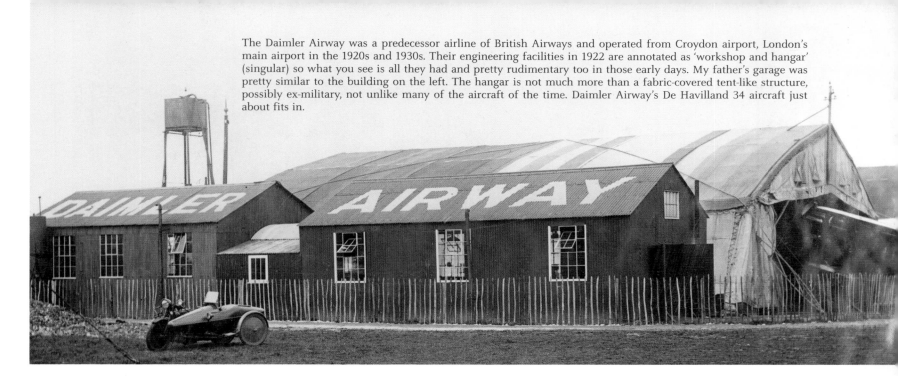

The Daimler Airway was a predecessor airline of British Airways and operated from Croydon airport, London's main airport in the 1920s and 1930s. Their engineering facilities in 1922 are annotated as 'workshop and hangar' (singular) so what you see is all they had and pretty rudimentary too in those early days. My father's garage was pretty similar to the building on the left. The hangar is not much more than a fabric-covered tent-like structure, possibly ex-military, not unlike many of the aircraft of the time. Daimler Airway's De Havilland 34 aircraft just about fits in.

Opposite, main: This is the main engineering hall of BOAC's main hangar complex TBA at Heathrow airport circa the mid 1960s. There was a multitude of different activities taking place in the hall making a variety of different, and sometimes very loud, noises, all echoing around the cavernous structure. British Airways Engineering does it very differently today with many activities separated into their own workshops or even separate businesses. My office was half way down on the right hand side overhanging the sheet metal shop with the 'tin bashers' banging away below. Opposite my window on the far side was the dinghy bay, a generally quiet area except when on occasion a dinghy on test blew up giving me and others close by a large adrenalin shot!

Opposite, inset: These engineers, clearly posing as they pretend to push up a main landing gear into its wheel-bay, an impossible task by the way, rather epitomise the 'dirty overalls and spanners' impression of many of the engineering ranks in the 1960s and 1970s, and possibly later. The dirty overalls are definitely not posed and getting mucky was part of the job. The aircraft is jacked-up so I suspect a wheel change was underway, definitely a mucky job. Quite why white overalls were standard kit I'll never know, the colour was certainly impractical. For many years, especially pre-War, engineers wore dark blue overalls and supervisors wore white long coats as a mark of identity and rank. White for everyone would come later. My father wore dark blue overalls in the 1940s so while the dirt did not go away it was better hidden. Maybe it was a mark of a higher skilled man to wear white with lower skills wearing darker colours. The supervisor here is the man on the left encouraging a hernia and wearing a peaked cap, definitely a mark of rank, and two stripes on his epaulettes.

This spread: The flight engineers of later years had much more room to do their job and in the comfort of an enclosed cockpit, but its complexity had increased enormously by the early 1950s. Civilian airliners benefitted from a wide range of new technology and the larger ones were multi-engine requiring a multiplicity of back-up systems. The flight engineer's principal role then was to support the pilots in operating the aircraft in flight so no hopping outside for external adjustments except at the airport of destination. In this image of a Lockheed Constellation aircraft from the early 1950s the flight engineer sits on the right at his instrument panel behind the two pilots. There would also have been a radio operator/navigator on the left of the cockpit just out of the picture. It all looks very complex and it was. Nowadays, it is still complex but computer screens and avionics systems do away with the need for many of the dialled instruments so no need for flight engineers and radio/navigators in the 21st century aircraft's cockpit.

Opposite: The Daimler Airway workshop had its engine test shop at the far end. The noise must have been horrendous for the engineers at the other end working on the engines before they were tested. I've not seen any images of engineers wearing ear defenders, but then excess noise was not generally considered a health hazard in industrial premises in those early days. The two parallel tracks in the floor are railway lines used to move engines on their 'dolleys' from the maintenance area to the test stand. The little cast iron stove in the middle left of the image is an example of an almost Dickensian method, by today's standards, of keeping the work environment warm in the winter months.

This page: Tuning engines prior to testing was a highly skilled job. Given that aviation was not much more than twenty years old at the time the image was taken, the engines themselves look complex and quite sophisticated and are a good example of how war can drive advances in technology – the First World War having ended just four years before in 1918. Skilled men were needed to work on the engines and they must have been in demand by the many small, private UK airlines then operating. Many came from the fledgling Royal Air Force or were trained by the manufacturers in the immediate post-war years. White coats appeared to be standard wear for the engineer even in these early years, albeit they were not that practical and became dirty very quickly. They certainly appear to be a mark of a higher skilled person. Other men employed to do more physical work such as removing and pushing the engines around tended to be in cloth caps and manual work wear.

These images are annotated 'Testing plugs before flight'. I suspect there was a very good reason for needing to keep the engine running in order to test the plugs properly, but it looks a precarious operation with the propeller whizzing around less than one metre away from the engineer's head. The engineer in the cockpit seems to be in the safest place operating the throttle. Images from these early years, and even later into the 1960s, show a remarkable lack of safety measures by today's standards. There are no harnesses of any kind to prevent falls and I bet the engineer on the ladder putting in petrol was not ladder trained; at least he does not appear to be smoking!

Right, above and below: Inspecting the tail plane hinges was a lot less precarious provided one's fingers did not get trapped. I suspect that this task was a standard maintenance procedure and an important one. The airfields of the time were nothing more than grass strips and muddy and very bumpy in wet weather. Losing parts of the tail plane in-flight was to be avoided at all costs so regular inspections were a critical safety measure. There is a distinct lack of supporting equipment in these early images and much of the heavy lifting and moving appears to be by muscle power alone. Injuries must have been an accepted occupational hazard – moving an aircraft in thick mud even a short distance must have required exceptional strength from one man alone.

Next spread left: BOAC's modern engineering base at Heathrow airport near the Hatton Cross Piccadilly Line Underground station, starts to take shape in the early 1950s. Its new headquarters building (TBA) and hangar complex are only half built but already dwarf the more traditional large black hangars that can be seen about a half mile away that were erected shortly after Heathrow opened in 1946. There are also a number of much smaller buildings or 'sheds' in the middle left of the image and one of these was the BOAC brake shop with its very grimy interior. I really don't know now which one it was but it has long gone. The black hangars are long gone too but several existed right up to the late 1990s. While they no longer housed aircraft they made very good storage facilities.

Next spread right and following page: TBA has worked well for over sixty years and continues to do so having recently had a new lease of life. Many different types of aircraft from Lockheed Constellations in the 1950s to Concorde in the 2000s have fitted into TBA's four large hangars. The central engineering hall has changed beyond recognition, initially housing a multiplicity of different trades and activities in 1954 to housing British Airways' simulators in the 21st century. It says something for the vision of the designers of TBA all those years ago that the building still performs a very valuable function for today's operations.

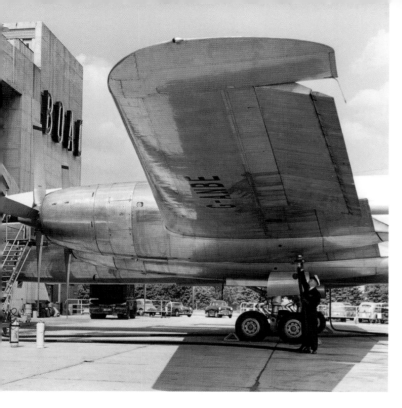

This page and overleaf: The massive size yet pleasing, almost art deco, design of the vast concrete spans and supporting pillars that front TBA's four main hangars or 'pens' as they were originally designated. Using the points of the compass to identify each hangar i.e North Pen on the north east corner of TBA, West Pen on the north west corner, South Pen on the south west corner and East Pen on the south east corner, this was a very different hangar in terms of size and design from anything else built at Heathrow and remains so to this day. I'd suggest the hangar facades are an exceptional example of industrial architecture, although three are now hidden away by later extensions. The East pen, however, was designated Grade: II and remains clearly visible.

This page and overleaf: BEA's main hangar at Heathrow was completed in 1953 ready for the airline's move of its operations from Northolt airport to Heathrow in 1954. It was a very different and rather contemporary design to TBA being built as a series of ten linked hangars or bays five on each side of the building with a series of workshops in-between. Not much concrete was visible and it was all steel, glass and cladding and very modern and suited BEA's purposes very well. Most of the aircraft in BEA's fleet returned to Heathrow each evening so any defects needing maintenance could be quickly carried out and the aircraft were fully serviceable to commence operations early the next day. The hangar complex did not house large numbers of offices as did TBA, which had been designed to bring together BOAC's commercial, operations and engineering facilities, one reason it was so large. BEA's hangar was almost pure engineering and their commercial people remained near Northolt in a converted school at Ruislip until its merger with BOAC in April 1974 to form British Airways.

Previous spread right and opposite: BEA's main hangar was initially known simply as 'Engineering Base, London Airport', changed to 'West Base' in later years and finally just 'TBE'. It was designed to house BEA's then new aircraft the Vickers Viscount and also the Airspeed Ambassador, often referred to as the Elizabethan.

Above right: BEA's original fleet of Vickers Viking aircraft were becoming outdated by the early 1950s and would be retired in 1954. Rather than tool-up for them to be maintained at its new engineering base at Heathrow, it made sense to continue to carry out maintenance at either BEA's Northolt base or Glasgow; these are two of the later and larger thirty-two-seat Vikings at Northolt undergoing a light maintenance check. In Northolt's early days fifteen or more Vikings would be serviced every night in No. 5 and No. 6 hangars.

Below right: Glasgow was an important UK destination for BEA and its Glasgow engineering base hangar was so useful it continued to be used for many years even after BEA moved to Heathrow; a British Airways Highlands and Islands Division British Aerospace Advance Turboprop (ATP) aircraft undergoes an engine check at the Glasgow hangar in 1993.

2

MAKE·DO·AND·MEND

In the very early days of commercial civil aviation the role of engineering was primarily about keeping on top of the unreliability of the aircraft of the time; the flimsy aircraft that were available broke easily and their engines were notoriously fickle and often stopped working. Not a pleasant thought several thousand feet up in the air – as someone said in later years 'there are no garages at 30,000 feet'. Imperial Airways, a predecessor airline of British Airways and the UK's first nationalised airline, came into existence in April 1924 following the merger of four small UK private airlines whose various fleets of aircraft and spare engines were old with many obsolete or already broken. In late 1924 Imperial had to write-off ninety of its inherited engines for just these reasons. Imperial's early services could often be late or cancelled for lack of serviceable aircraft, a situation that could not be allowed to continue without risking losing business, even on Imperial's monopoly routes where it was the sole airline operator.

Imperial's ground engineers must have operated from time-to-time on a make-do-and-mend basis, not too difficult a task with most aircraft being made of wood and canvas so relatively easily patched and glued back together. While there was close UK government oversight of the manufacture and maintenance of civil aircraft and there were manufacturers' maintenance manuals and spare parts to work with, there would probably have been many occasions when an engineer had to find his own way around fixing a problem, using his own tools, experience and common sense. I remember my father telling me about doing just that during the Second World War when he was on reserved occupation mending damaged military aircraft as well as building new ones from prepared kits at Hanworth aerodrome, a stone's throw from the new airport about to be built at Heathrow. All this was good experience for early peacetime services when civilian flying restarted in 1946; maybe that is why BOAC employed him that year, but it was hardly a basis for restructuring into a more precise, regulatory and technologically led maintenance organisation in later years.

We might be surprised today by some of the 'technology' used to get around an aircraft maintenance problem in the 1920s and 1930s, but if necessity is the mother of invention then some quite ingenious ideas were put into practice. The development of flying boats created maintenance challenges in their own right let alone from a purely operational perspective, but couple that (literally) by attaching two flying boats together to increase the range of the smaller one, was a wild idea (that worked) only revisited in modern times with the space shuttle concept. In-flight refuelling was another idea for increasing a flying boat's operational range that nowadays is a commonplace arrangement

Previous spread, left: Croydon Airport had become London's main civil airport by the time Imperial Airways was established in 1924. Some of the buildings seen at the top of the main image were part of Imperial's engineering base and were not that much different from the small workshops and hangars of a few years before. This is a very early image showing other buildings used as administrative offices or check-in facilities and one of Imperial's Handley Page W8 aircraft waiting for passengers to be embarked. The W8 was a rather flimsy bi-plane and was a conversion of an ex-military First World War type that was not expected to have a long service life. It was a twin-engine aircraft and having two engines was not just about having enough power to fly the aeroplane but a way of reducing the risk if one engine failed. Should that occur then the second engine could keep the aircraft flying – the manufacturers even advertised that their twin-engine aircraft could 'fly for many miles on one engine only', a reassurance for passengers and for Imperial's reliability as it meant there was a fair chance the aircraft would keep on time. These larger aircraft did not try to land in convenient fields for temporary repairs to be carried out (they were getting a bit too large for that sort of thing) but relied on Imperial's ground engineers to, hopefully, fix the engine at the aircraft's next scheduled stopping point.

Left: We know that Imperial's aircraft fleet comprised just fifteen usable aircraft once it had sorted out the best aircraft from the fleets of its four predecessor companies. This was one that did not make the cut as it had been used as an engine test bed and was scrapped. The unreliability of aircraft engines had quickly led to the development of leasing arrangements with the engine manufacturers rather than an expensive purchase. This not only reduced the financial burden on the UK Exchequer, who would otherwise have to guarantee any purchases as Imperial was reliant on UK government support, but also encourage the manufacturers to invest in research and development to improve their products.

but only for military aircraft. Its use for a civilian airliner in the 1930s, let alone the almost theatrical method of transferring fuel, can only be imagined. It also worked and while it was the aircraft designers and manufacturers who came up with most of these concepts, it was the engineers that put them into day-to-day practice and made them work.

New equipment was also being rapidly developed on board aircraft in these early years. Science was emerging behind new maintenance procedures and practices as centralization and standardization began to emerge. Radios began to come into use and, later, direction finding equipment allowing all-weather operations. In-flight catering and the introduction of galleys and boilers created their own headaches for engineers to solve.

When BOAC started commercial operations in May 1946 it had 169 aircraft in its fleet of nineteen different types with nineteen different engines. BOAC had been formed in 1940 from an amalgamation of Imperial Airways and a small private UK airline, British Airways Ltd, and had flown throughout the Second World War supporting the war effort under the control of the UK Ministry of Aviation. At the end it was pretty well worn out and had to rebuild itself from its wartime operation into a commercial organisation in a very short time. Resources were scarce and scattered about, not least in its engineering department with facilities as far away as Dorval in Canada to Filton, near Bristol and Southampton. Centralising its facilities, not least in hangars and spares, ideally to the new London airport, now to be its new main operating base, would take several years yet to complete. Most of BOAC's aircraft and engines were a collection of largely obsolescent hand-me downs from Imperial Airways and war-time conversions. My father had worked on a few of these during the war as no doubt had many of his new colleagues employed from reserved occupations or demobbed from the armed forces. A

notable ex-military type was the Avro Lancastrian, in its military form more used to dropping bombs than dropping off passengers, and the Douglas DC3. BOAC's sister nationalised company, BEA, had only been formed in August 1946 and was not in much better shape given it had also inherited several obsolescent aircraft types including several German Junkers Ju52s taken as war reparations, a type previously used, amongst many other things, for dropping parachutists. I don't believe BOAC's engineers had to remove any bomb racks before installing any seats or BEA's engineers any parachute harnesses, but what is certain is that those engineers probably would have had a lot to do to modify those respective aircraft and other types before they could be used for service. To BOAC's and BEA's engineers this would not have been at all unusual. Modifying new aircraft coming into service or during their operational lives to suit an airline's particular operational or commercial requirements, had been in practice since civil aviation began and still applies today. The next few years, however, would see BOAC's and BEA's engineers having to work very hard to keep what aircraft they had still flying and to an acceptable operational schedule and at the same time reorganise themselves also to introduce new aircraft into their fleets. The phrase 're-engineering engineering' was never more apt than in the years from 1946 to the mid 1950s.

This brings us to a central issue that airline engineers have had to tackle on a quite frequent basis – the introduction of new types of aircraft into an airline's fleet. This could happen in time spans of just a few years at a time, such as in the late 1940s and early 1950s, as new aircraft rapidly came off the manufacturers' production lines and almost as rapidly were sold on. Up to the late 1960s there were a number of different UK airline manufacturers all competing for BOAC's and BEA's business. With rapidly advancing technology they were all trying to get one step ahead with new aircraft types and improved propulsion systems. As the two nationalised airlines of the UK both were expected to 'buy British', primarily as a means to support the UK aircraft manufacturing sector. There were exceptions to this 'rule', however. There had to be because there were several occasions where the UK manufacturers just could not deliver on what BOAC needed for its long-haul services; it needed long-range, four-engine aircraft that could be competitive with those used by its international rivals, particularly the main US airlines Pan American and Trans World Airlines on the important Atlantic routes. BEA, on the other hand, generally had a better time of it as the UK airline manufacturers were quite good at producing smaller aircraft, some such as the Vickers Viscount with 'turbo-prop' part 'jet', part propeller, driven engines that were so good even the US airlines bought them.

New aircraft, new systems, new maintenance practices and procedures had to be learnt and applied quickly as these aircraft came into BOAC's and BEA's fleets. A critical first step, however, was the need to satisfy the UK aviation regulatory authority of the time, the Air Registration Board, that both companies had engineering and operational organisations with facilities and suitably licensed engineers capable of maintaining each aircraft so they were fit to fly and eligible to be issued with a certificate of airworthiness. Engineering an airline is therefore much more than simply maintaining aircraft. Effective maintenance is the end result but it is the product of a close and continuing co-ordination and co-operation between the airline concerned, its aircraft (and component) manufacturers, particularly for engines, and the national aviation regulatory authority (for the UK that is now the Civil Aviation Authority) and all within a tightly controlled regulatory framework that governs international civil aviation. As an example, in 1949 and because of the imminent delivery of ten Boeing Stratocruisers, BOAC established

a permanent representative's office at Boeing's US manufacturing facility in Seattle to ensure the aircraft met BOAC's and the UK Air Registration Board's requirements. Because of the growing importance of Boeing aircraft to BOAC's operations throughout the 1950s and 1960s, and into the 1970s with British Airways, the practice continued and continues to operate today. Similarly for its Airbus aircraft, British Airways engineering specialists maintain a close relationship with that manufacturer, but their bases at Toulouse in France or Erding in Germany are much easier locations to visit for a day. Rolls Royce and British Airways have also worked closely for many decades given Rolls Royce engines were and continue to be widely used across the airline's Boeing and Airbus fleets and continues a long tradition established by BOAC and BEA as long ago as the 1940s.

Simply put, this essential co-operation, even between competing airlines, is in order for aircraft to remain safe and efficient to operate. Approved maintenance programmes are jointly developed for each aircraft type and also for their major components such as engines and landing gear assemblies. These cover an aircraft's daily or weekly checks and inspections, known as line maintenance, to 'light' or 'heavy' checks after set numbers of hours flown or months operated. The size and shape of an airline's engineering organisation must therefore constantly evolve to reflect the many and complex requirements of their respective aircraft fleet's maintenance requirements and changing technology, probably the most significant being the rapidly improving reliability of 21st-century aircraft and their components, particularly engines.

An aircraft's engines may appear to be the most complex component of any aeroplane, but cockpit systems, particularly 21st-century avionics, beat that by a long way as the jet engine is a relatively simple arrangement. It was not always like that with piston engine propeller driven aircraft from

This page: We don't know how Imperial's management sorted out what its engineering workforce structure should be as it took on what were four different engineering groups from its four predecessor companies, but an early operational decision was to adopt the same policy as Handley Page and opt to fly multi-engine aircraft. This requirement was linked to Imperial needing larger, more powerful and reliable aircraft if it was to have any chance of success in developing the Empire air routes for which it had been established. Imperial specified a three engine layout for its new aircraft that had yet to be designed and built. The Handley Page Argosy and De Havilland 66 were the two aircraft finally chosen and they became the main stays of Imperial's fleet throughout the 1920s. As the Empire air routes developed, it quickly became necessary for engineers to be posted overseas along the routes. They would then be in position to provide any engineering services required as in these early years there were very few local resources that could be utilised; it would be several decades yet and another World War before the possibility of outsourcing overseas engineering became a realistic option.

Next spread: Facilities for overseas engineering in the 1920s could be quite basic. A hangar to protect the aircraft, and its ground engineers, from the elements – as diverse as very high temperatures, sandstorms and even heavy rain and hailstorms – was very important. These images are of an Imperial Airways Handley Page HP42 at Karachi. Its hangar was originally designed for the UK's ill-fated R101 airship and following its destruction was used by Imperial to shelter its aircraft from the worst of the outside weather. With working temperatures inside often well over 40 degrees centigrade and the roof so high it was said to generate its own clouds during the monsoon season; it's questionable how much protection it gave to Imperial's engineers.

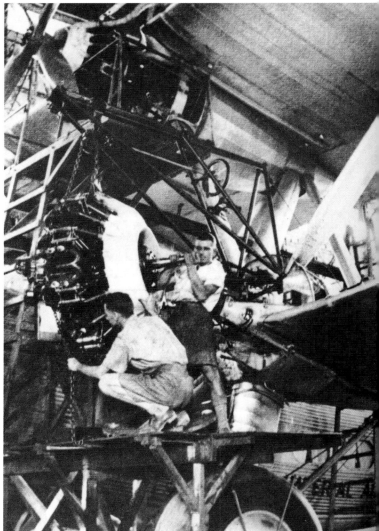

the 1940s and 1950s becoming increasingly complicated. Advancing technology during the Second World War had increased the power of piston engines enormously, but in doing so had also considerably increased their complexity. In the early 1950s the introduction of the 'turbo-prop' engine, basically a piston driven aircraft engine that utilised its exhaust gases for additional thrust, gave the best in terms of power output with a relatively small increase in complexity. Pure piston engines were, however, reaching the limit of their development potential and the introduction of the first civilian jet aircraft, the De Havilland Comet 1 in 1952, as good as ensured they were on the glide path to obsolescence. Just like Concorde twenty five years later, the introduction of the Comet 1 with its four powerful De Havilland Ghost jet engines was both ground breaking and record breaking. At its operating altitude the Comet 1 flew at around 500 miles per hour, broadly the speed of commercial jet aircraft today and over 200 miles per hour faster than the piston engine propeller aircraft of the time. It effectively shortened journey times into hours rather than days but its promise was short-lived. After two relatively trouble-free years the Comet 1s suffered a major structural failure and were grounded. This was no fault of BOAC's engineers but an example of technology moving ahead of an understanding of the science behind it. Checking for metal fatigue is now well understood and has become a very important part of any airline maintenance programme and a very good example of engineering science in practice.

For five years the promise of the jet engine remained just that, a tantalising promise. BOAC had to continue to operate and engineer older, piston engine propeller aircraft such as the Canadair Argonaut and continued to use the Boeing Stratocruiser for more years than it was originally intended. Yet another new aircraft, the Douglas DC7C, was also ordered to help fill the gap in the transatlantic fleet pending the later than expected introduction of the new and much heralded, long-range Bristol Britannia 102. Engineers had to as good as forget their Comet training and retrain to maintain aircraft they had never expected to work on again as well as continue for much longer the labour intensive maintenance and operation of the older, often unreliable piston engine. This complexity of aircraft types also exacerbated BOAC's failure to keep its engineering costs in line with its major competitors. BOAC's maintenance organisation was split by type into aircraft fleets so any new aircraft would have its own engineers, tooling and spares but often share the same hangar space as other aircraft type fleets. This inevitably led to duplication of both men and materials and BOAC's cost per man hour was considerably higher as a result. It would be several years before the issue was tackled and engineering re-organised into heavy maintenance and line maintenance covering all fleet types.

All this was just catch-up for the main event, however. The later 1950s did this time represent a real turning point in civil aircraft engineering history as in 1958 the re-engineered jet-powered Comet 4 was introduced on BOAC's all-important North Atlantic route narrowly beating Pan American's introduction of the larger, also jet-powered, Boeing 707. The speed advantage of jet travel was now here to stay and the late introduction of the turbo-prop Britannia a year earlier in 1957, was a good example of being too late in the game. The aircraft did have good economics and passenger appeal and actually lasted into the 1970s with a number of UK charter airlines, but it could not match the commercial advantage of jet travel. Recognising this, BOAC had sold its Britannia fleet by the mid-1960s.

Following the Comet and 707 new jet engine aircraft of different types were being ordered by all the larger airlines, initially mainly for long-haul, intercontinental travel. Engineering maintenance capabilities were also rapidly being expanded to cater for jet aircraft both at BOAC and, a little later, at BEA. The jet was clearly the future aircraft but jet speed over the much shorter European and UK domestic routes was not such a major advantage. BEA eventually bought the shorter-range Comet 4B and later the De Havilland Trident and BAC 1-11 but continued to use several propeller driven aircraft right up to its merger with BOAC in 1974, notably the by then quite old Vickers Viscount. From an engineering perspective, it was the improved reliability of the jet engine over the piston-engine that was the major prize in this contest. This meant longer intervals between major maintenance and longer periods 'on wing', a considerable improvement in man-hours and costs. These early improvements are, however, dwarfed in comparison to more modern jet engine types such as the Rolls Royce RB 211-524G engines used on British Airways' current 747 - 400 aircraft. The Comet 4 Avon engine was doing well to get 3,000 hours life 'on wing' between overhauls, piston engines were only around 2,000 hours. By contrast the RB 211-524G averages between 38,000 and 40,000 hours 'on wing' for up to nine years running for twelve hours a day on average. That is a phenomenal improvement and an indicator of the huge advances in aircraft and engine performance and reliability in the 21st century. It's the main reason why airlines continue to buy new aircraft to gain the latest advantages in both operational and cost performance. Back in the 1960s many charter and third level scheduled airlines bought the larger airlines' second-hand aircraft and BOAC's Britannia's soldiered on for many years as charter aircraft.

By the early 1960s the escalating costs of operating more modern aircraft types forced all airlines to examine the economics of their maintenance activities as a whole. Using analytics to measure aircraft failure patterns, it could be proven that many scheduled maintenance tasks were not necessarily contributing to the safety or economics of the aircraft at all with some tasks even detrimental to reliability. This analytical approach eventually led to a considerable overall reduction in scheduled maintenance with significantly longer periods between checks and a decrease in scheduled tasks. Using the 747 as a model, a formal logical analysis process was developed by BOAC's engineers in collaboration with Boeing further accelerating what was to become the changing face of modern aircraft scheduled maintenance; within only four years the 747's intermediate check time intervals were increased from 600 hours to 2,400 hours. This was a significant improvement providing not only major cost savings but a better use of overall resources and an actual improvement in safety standards. By comparison, it had taken ten years since its introduction in the mid 1960s for the intervals between checks on BOAC's much smaller Vickers VC10 aircraft to increase from 300 hours to 1,500 hours. Analytical techniques had also highlighted that an airline's ability to react to technical problems was just as important as how it carried out scheduled maintenance. 'Condition monitoring' became the code name for measuring reaction ability based on the effectiveness of communication methodology within the airline's engineering operation of information on the state of an aircraft and its parts. This process now underpins all aircraft movements across the British Airways' fleet – an almost instant analysis wherever the aircraft might be, whether in the air or on the ground.

There's an interesting statement in the BEA annual accounts from the mid-1960s that opines, "The greater complexity and high cost of modern aircraft demand a higher degree of engineering perfection than was the case with older and cheaper aircraft". This might suggest a lesser attention to detail in earlier years, but what it really meant was the need for much closer attention to improving not only airframe and engine

lives and reliability, and associated costs, but also components, not least because spares and materials were predicted to become the biggest future annual cost for the airline, even above labour costs; expensive components, for example, were often routinely replaced at an aircraft's set check intervals regardless of whether the component's life had been reached. This had to change and analytical techniques would become a major factor in making this happen. BOAC had come to a similar conclusion and was also looking at achieving effective product support and adequate component warranties, something previously never effectively dealt with. With ever larger aircraft coming into both BOAC's and BEA's fleets, costs were also escalating in the provision of ground support for the larger jets, let alone the need for additional maintenance facilities such as new hangars. BOAC was also beginning to worry about whether the new supersonic aircraft that were being developed would effectively make most of their current subsonic fleets obsolete before the end of their economic lives, just as had happened to both the Britannia and DC7C. The cost of engineering was rapidly taking a higher profile in both BOAC and BEA and would became increasingly critical in the next decade and beyond.

Left: This image is a good example of Imperial's engineers finding a way around a problem using what was to hand, common-sense and a fair bit of muscle. Maybe this was an approved method at the time of lifting a damaged aircraft using baulks of timber and a block and tackle; it did work but the safety of the men climbing the wooden tripod is certainly questionable. Nowadays, air bags and jacks would be more appropriate. This Handley Page W8 of Imperial's fleet had come in to land too fast, hit a bump in the runway and its undercarriage had collapsed. It is a good example of the old engineering complaint that pilots break aircraft so engineers can fix them, and they did with this one, which flew again.

Opposite: A few years later, in 1932, one of Imperial's Handley Page HP42 aircraft lost its undercarriage in a similar incident by hitting a slightly raised concrete tunnel at Hanworth aerodrome. It also was fixed and flew again but this time a more modern way of lifting the aircraft was used courtesy of a Beck & Pollitzer crane. The wooden beams and block and tackle of a few years before had been consigned to history as aircraft became larger and heavier and more modern methods of aircraft recovery and maintenance were developed.

Opposite: We don't quite know what is happening to this Imperial Airways' Short L.17 landplane Scylla in the mid 1930s. It may just be undergoing major maintenance to replace the top wing structure and engine cowlings, possibly following a heavy landing; a multiplicity of solid timber supports are taking the strain of what appears to be a difficult engineering task for Imperial's engineers.

Above: This image of an Imperial Airways' Handley Page HP42 aircraft undertaking a major overhaul in 1933 at Imperial's Croydon airport base is a good example of the development of more modern methods of aircraft maintenance that were gradually being put into practice during the 1930s. Simple staging has been introduced and erected around the aircraft to make it much easier and safer working at heights but personal safety equipment is noticeably lacking; standing on an oil drum without a face mask close to the spraying of what looks like paint, let alone balancing on some sort of tripod steps without safety rails or a harness, would be illegal today. Incidents were not uncommon, some serious, which ultimately would lead to the beginnings of a much more safety-conscious environment.

The Handley Page HP42 aircraft fuselage was largely metal skinned requiring a new category of metal workers to be employed, but there remained a high proportion of wood and fabric construction, particularly on the wings, that still required engineers to be employed that understood the working of that material and its technology. Imperial also continued to employ a high proportion of airframe riggers competent to maintain the integrity of the many bi-plane aircraft that Imperial still used right up to the beginning of the Second World War. The bi-plane was now old technology and rapidly becoming obsolete as more modern all-metal monoplane aircraft came into service with Imperial's competitors.

Above right: Imperial Airways did operate a few monoplanes including the De Havilland Frobisher shown in this image of the aircraft undergoing maintenance in Imperial's Croydon hangars. De Havilland had originally named the aircraft the Albatross but, understandably, Imperial decided to rename it. The Frobisher was certainly no albatross but very streamlined and fast breaking several air speed records in its short career; it came into service in the late 1930s and was used during the Second World War when the only six produced were lost. Interestingly, while a modern monoplane design, it was largely made of wood and wood veneer and needed special care and attention in bad weather to protect it to limit the risk of water damage, something else for Imperial's engineers to deal with. The Frobisher came from the same manufacturer as the wartime De Havilland Mosquito fighter-bomber that was also of wooden construction; but then its life expectation was a matter of weeks not years.

Below right: This Frobisher is undergoing pre-flight maintenance at Croydon before its next scheduled service. Checking the undercarriage and wing control surfaces is a daily set of tasks for any operational aircraft even into the 21st century.

Next spread left: The name 'flying boats' was a misnomer as they were not boats that flew but real aircraft that took-off and landed on water. This image of a Supermarine Sea Eagle flying boat at Southampton Water in the early 1920s does, however, certainly look like a boat with wings and the way it is being handled, especially by the man at the bow, or should it be the nose, certainly suggests that. Operating flying boats had their own engineering challenges let alone operational difficulties, which this image certainly brings out. Safeguarding the passengers was clearly very important and several men, engineers, dock hands, we can't be sure who they are, cluster around the gangplank leading on to the jetty. The passengers are VIPs, Sir Sefton Branker the then UK Minister of Aviation, and an unidentified lady passenger. Sir Sefton had been on a trial flight of the flying boat so great care was being taken to keep him and the other passenger safe. This particular flying boat was operated by British Marine Air Navigation Company between Southampton and Guernsey in the Channel Islands; BMAN was taken over by Imperial Airways in 1924 but by then they had already lost the Sea Eagle seen in this image as it was damaged at Alderney in 1923 and not repaired.

Another example of the challenges faced in maintaining flying boats. This late 1940s BOAC converted Short Sunderland is undergoing a pre-service check that could be done while afloat. The issues for the engineers was not just one of carrying out the service checks but doing so safely and not falling in the water, a hazardous business especially if done at night in freezing conditions when hessian sacks were used for a better grip when walking on the aircraft's icy wing surfaces.

Left: We think this wheel had just been taken off a Solent flying boat after it had been wheeled down the jetty in to the water and floated. The engineer has then disconnected the undercarriage leg and is dragging it back to the jetty with the wheel assembly partly afloat due to the air in the tyre. There were clearly some risks associated with this method of recovering equipment, not least in the safety of the engineer who appears to be in some sort of life preserver but with no apparent safety harness and rope between him and dry land.

Below: In the 1920s, to overcome some of the difficulties of winching its flying boats out of the water for maintenance, Imperial experimented with the idea of a floating dock. The RAF had originally trialled the dock, which could be partially submerged to allow the aircraft to enter, and then be winched and manhandled into position and secured before the dock was drained. This image shows an Imperial Airways Short Scipio flying boat secured in the dock during trials at Southampton.

This page, overleaf left: Maia was the name given to a modified Short 'C' Class flying boat of Imperial Airways that had special fittings made for its upper fuselage to connect a smaller Short S.20 flying boat named Mercury on top. The idea came from Imperial's own engineers so that longer range operations could be carried out for carrying mail rather than passengers, particularly to allow the first transatlantic services to begin in the late 1930s. The idea was considerably ahead of its time and allowed Maia to take-off carrying Mercury thereby saving the smaller aircraft's fuel for the Atlantic crossing. Several speed records were subsequently established across the Atlantic and, later, to South Africa, but rapidly advancing technology in flying boat design and more powerful engines enabled the first regular Atlantic mail flights to begin by a single flying boat in August 1939.

In-flight refuelling for its large Short 'C' Class flying boats was another idea ahead of its time pioneered by Imperial Airways for civilian operations. Given the flying boats were propeller-driven, the refuelling pipe of the tanker aircraft could not be trailed ahead of the flying boat without a serious risk of fouling its propellers. Imperial got around the problem by the flying boat pilot flying under the refuelling pipe and manoeuvring his aircraft to fly just ahead of it close to the rear of the fuselage. The flight engineer then crawled into the tail of the flying boat, opened a small hatch and hooked the refuelling pipe so it could be drawn into the aircraft and connected to its fuel supply pipe, removing it the same way. The intention was to use this method to increase the range of Imperial's mail carrying flying boats, but, as with the Mercury and Maia idea, technology caught up and the arrangement was not put into practice.

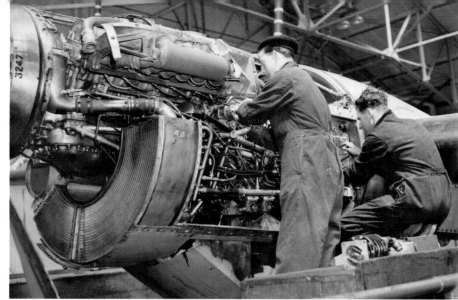

Above left and below left: What we now call avionics rapidly developed during the Second World War with increasingly sophisticated (by the standards of the day) methods of direction finding and communication. Radios were becoming standard equipment on passenger aircraft during the 1930s and together with more complex instrumentation, all needing to be overhauled and calibrated, the calls on aircraft engineers became ever more demanding. Not so sophisticated but with their own technicalities (and frustrations) to master were in the aircraft galleys as boilers and ovens became installed to cater for higher levels of passenger comfort; these early galleys would not have looked out of place in a kitchen at home, but coupled with the stresses and strains of an operational aircraft their equipment often broke or was just temperamental.

Above and opposite: Many of BOAC's engineers would have been familiar with the Rolls Royce Merlin engine that was widely used during the Second World War to power both fighter aircraft and bombers, notably the Supermarine Spitfire and Avro Lancaster. It was such a good engine it was also used after the war on civilian aircraft such as the Avro Lancastrian, a conversion of the Lancaster bomber, and the Canadair Argonaut. I know my father worked on them both during and after the war and these images show BOAC engineers working on an Argonaut and its Merlin engines in one of BOAC's early black maintenance hangars at Heathrow before TBA was completed.

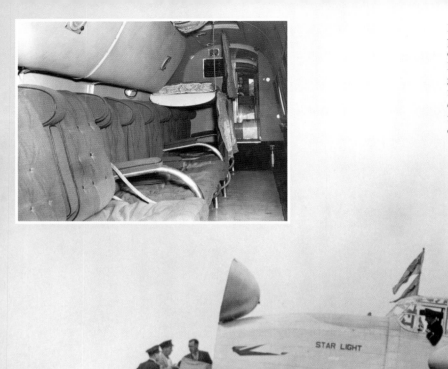

It was a Lancastrian operated by British South American Airways, another predecessor airline of British Airways, which made the first flight out of the newly opened London Airport (Heathrow) on 1 January 1946. These images show a BSAA Lancastrian being loaded before departure with what would have been the front gun turret (on a Lancaster) conveniently modified into a cargo compartment. The main modification, however, was to the aircraft's interior with a conversion to a one-abreast thirteen seat configuration, all that could be fitted into the narrow fuselage. BOAC's Lancastrian's only managed nine seats side-by-side, although with overhead bunks to use on their sixty-three-hour flights to Sydney, Australia.

This is one of BEA's ex-German Luftwaffe Junkers Ju52 transport aircraft from a fleet of ten taken over from Railway Air Services in 1947. They were, apparently, easy to fly and maintain except that useable spare parts were in short supply. BEA spent £12,500 modifying the interior of each aircraft for passenger use – rather a waste of £125,000 as the aircraft were all scrapped a year later. This was, however, the best decision, although they probably should have been scrapped a lot earlier. The Ju52 was a 1930s design with 1930s peculiarities and BEA had enough problems on its plate juggling maintenance operations with its other, more modern, fleet aircraft without having to search high and low for usable Ju52 spares, many of which were sub-standard war-time stock.

The Boeing Stratocruiser was BOAC's first commercial Boeing aircraft introduced in 1949 and a world away from the types it had operated pre-war being the height of passenger comfort and luxury. In engineering terms it represented the latest in technology but it was a maintenance intensive aircraft and prone to engine failures. Its four Pratt & Whitney Wasp Major piston engines were very powerful but complex. Each engine had twenty eight cylinders arranged in a spiral to enhance cooling and had two spark plugs to each cylinder so fifty six in total; together with the associated wiring looms and electrical connections, each one of which could generate a fault, it was not an easy aircraft to maintain especially in poor weather outside of a hangar.

The Airspeed Ambassador, designated the 'Elizabethan' by BEA, was the first of BEA's new, modern and pressurized airliners but the last to be powered by an old technology piston engine, the Bristol Centaurus 661. Introduced in 1952 it was popular with passengers but quickly outclassed by the new turbo-prop Vickers Viscount that BEA introduced only a year later. The Centaurus 661 was a powerful engine and easy to access for maintenance but it was very complicated and unreliable as a result; it was so unreliable that BEA even kept one of its Bristol freighter aircraft on stand-by so it could fly out a spare engine to rescue any Elizabethan aircraft stranded abroad with engine failure. By mid 1958, the aircraft was retired after only six years in service and sold on.

Opposite: BEA introduced the Vickers Viscount in 1953 and they were still flying but as part of the British Airways' fleet right up to 1982, a phenomenal record of reliability and solid service. Their four Rolls Royce Dart turbo-prop engines were complex but reliable and not that difficult to maintain. They were the last of the old propeller-driven aircraft in the British Airways fleet but still capable of producing an economic return versus jet aircraft when used on the shorter Highlands & Islands services in Scotland. The images are of Viscounts in BEA's new maintenance hangar at Heathrow in the mid 1950s undergoing various checks. The engine check is being closely watched by a supervisor in his white coat. Close supervision has always been a pre-requisite of aircraft maintenance right up to the beginnings of the 21st century when self-certification became a standard practice in British Airways.

Right: The De Havilland Comet 1 marked the beginning of the civil aviation jet age in 1952. Despite its withdrawal from service after only two years there would be no change in the move forward to develop a safe, civilian jet aircraft but just a delay to refine the original design. Re-engineered as the Comet 4 it re-entered service in 1958, this time with wing tanks to give it an extended range to fly the commercially important North Atlantic routes. Pictured in TBA's East Pen hangar on its delivery day, it is surrounded by propeller driven aircraft which it was rapidly to supersede with its superior speed, a Douglas DC7C in the far background and a Bristol Britannia in the foreground.

This spread: Technical Block 'B', or TBB as it became known, was constructed in the mid 1950s to carry out minor maintenance on the Bristol Britannia. Looking a bit like a double-sided conservatory tacked on to a central workshop and stores, TBB was originally called the 'Wing Hangar' because it was designed to allow most of the fuselage and the wing and engine surfaces of the Britannia to be protected but for the large and high tail to remain exposed; that meant the hangar roofline could be lower and save costs. It was said that should any engineer upset his supervisor then he risked being sent out in the rain and cold to work on the tail of the aircraft! Britannia major checks would continue to be carried out in TBA.

Due to the expected introduction of its next new aircraft, the turbo-prop Vickers Vanguard, in 1960 and further orders of the very successful Viscount, BEA had to build another new hangar next to and almost a mirror image in size and shape to its original main hangar at Heathrow. With ten bays and a taller roof line, it subsequently was designated 'East Base' (with the original hangar as 'West Base') and then 'TBD'. At one point the hangar was even used to maintain Concorde but it was demolished in 2009 to make way for a larger parking area for British Airways' aircraft.

This page: During the early 1940s to protect it from the risk of bombing, BOAC's engine overhaul facility at London's Croydon Airport was moved to Treforest in South Wales. Following the end of the Second World War and in order to save external engineering costs, BOAC decided to bring more of its engine overhaul requirements in-house and by the mid-1950s Treforest had grown into a substantial company maintaining both BOAC and other airlines' engines. It continued to expand in later years but was eventually sold by British Airways to General Electric Aircraft Engines (GE).

Overleaf: A Comet 4's Avon Mk524B jet engine stripped down into its constituent parts at BOAC's engine overhaul facility at Treforest, South Wales. Simply put, it's not much more than a box of bits that belie the powerful engine that it would become on construction. Powerful but relatively simple compared to the piston and even turbo-prop engines from only a few years earlier, the Avon was the forerunner of the very powerful jet engines that power 21st-century jet aircraft.

◄ B·O·A·C ►

REPAIR FACTORIES·TREFOREST

THE FIRST

AVON MK 524 B

TO REACH 2900 HOURS

BETWEEN OVERHAULS

The Comet 4B was BEA's first jet aircraft introduced in 1959. A version of BOAC's Comet 4 it did not have the long-range wing fuel tanks and was instantly recognizable because of that. It brought BEA into the jet age but it would be outclassed by the Hawker Siddeley Trident expected to be delivered within only five years. Unlike the Trident, which was a good enough aircraft to be developed into three variants over its fifteen year life, the Comet was a design based on 1950s technology and difficult to make bigger or take more powerful engines. BOAC, and BEA, recognised that good aeroplanes have always been capable of development into even better ones by the exploitation of technical advances to keep abreast of developing requirements. The Trident allowed that, the Comet 4B did not and was eventually relegated to charter work at BEA's subsidiary company BEA Airtours.

Above left: A year after the introduction of the Comet 4B, three more new aircraft types were delivered to BEA in 1960. The four engine turbo-prop Vickers Vanguard, Hawker Siddeley Argosy 101 and Handley Page Herald. The Vanguard was another expensive aircraft to introduce with a considerable amount of engineering work required pre-delivery let alone the need for a new hangar because of the higher tail fin. The aircraft also proved to be less than reliable, especially its new Rolls Royce Tyne engines, which at one time had so many failures some aircraft were grounded due to a shortage of spare engines.

Above middle: The Argosy 101 was a major improvement on BEA's very old DC3 freighter conversion, but it was expensive to maintain given only three were bought and the cost per aircraft was therefore high. During the 1960s there was a growing demand for air cargo and a cargo terminal was built at Heathrow to be shared by BEA and BOAC. The Argosy, however, did not have the capacity to meet the demand and were sold on after a few years following the conversion of several of BEA's Vanguards into large capacity freighters and renamed the Merchantman. This constant introduction of new fleet aircraft that lasted only a few years was a considerable cost burden to BEA well into the 1960s until their Trident and BAC1-11 fleets had become established and the Argosies, Heralds and Comet 4Bs had gone. Of BEA's legacy aircraft only the Vickers Viscount soldiered on, its reliability and cost effectiveness well established and its maintenance programmes finely tuned; in the early 1960s the Viscount's Dart 510 turbo-prop engines had the longest life on-wing of any aero-engine world-wide.

Above right: The Herald, also with only three aircraft flown, was another example of BEA being forced to support the UK aircraft industry. Its small fleet of three were actually leased from the UK's Ministry of Supply, not bought, and returned after only five years. The aircraft in this image, G-APWC, was modified by BEA to carry long-range fuel tanks and used by the Duke of Edinburgh who flew the aircraft extensively during a Handley Page promotional tour of South America in 1962.

This page and overleaf: BOAC's selection of the Boeing 707- 420 from a range of competing US and British future aircraft types had been carefully thought through and evaluated, although the British government disagreed and preferred the British Bristol Type 430, a turbo-prop almost 100 miles an hour slower. BOAC managed to win the argument and an initial order of fifteen was placed in late 1956 at a cost of £43,428,000, a considerable sum in mid-1950s money. The order specified Rolls Royce Conway engines and this marriage of Boeing airframe to Rolls Royce engines proved an ideal combination. Introduced on the North Atlantic routes in 1960, the 707 remained in operation up to and beyond the 1974 merger of BOAC and BEA. The 707 was another example of a good aeroplane being capable of further development with several different 707 variants being produced over the following years including a freighter version; in 1967/68, BOAC engineers extended the lives of twenty-two of the 707 fleet with wing modifications to provide an additional 20,000 hours of flying, broadly six extra years.

In the 1960s some maintenance tasks were nowhere near 21st-century standards. Working on the engine de-tuners in winter wearing your own coat and cloth cap while adjusting the Conway engines on a 707 was not a lot of fun! My father used to say that the Heathrow compass swing area and the de-tuners were the coldest places to be in the winter months.

This spread and overleaf: British government approval of BOAC's initial 707 order was contingent on their also ordering twenty British aircraft for their Eastern routes. BOAC's options were thus severely constrained, especially as no British manufacturer initially appeared interested; BOAC also would have preferred an all-707 fleet to standardise costs, not least engineering costs. After protracted negotiations with Vickers Armstrong and the British government, a design for a high-tail, 'clean wing' aircraft with four tail-mounted engines and named the Vickers VC10, was agreed. Thirty-five standard versions at £1.7m each and ten super versions at £2.55m each, were ordered. There were great expectations for these aircraft but strong suspicions that they would not be able to match the cost of operations of the 707, and so it proved. The VC10 was the first commercial 'clean wing' aircraft and gave good operating performance at 'hot and high' airfields where air density was lower and greater lift was needed. The standard VC10 was introduced in 1964 and, not unlike the 707, the marriage of airframe to Rolls Royce Conway engines produced a very well engineered aircraft capable of much more. During its planning for future supersonic operations, BOAC even circulated to aircraft manufacturers a specification for an up to 250 seat aircraft based on a development of the Super VC10.

A so-called 'fifth pod' on a VC10 encasing a spare engine. The fifth pod arrangement meant a spare engine could be carried externally rather than in the cargo hold and was a technique perfected on the 707 and VC10 for the first time in the 1960s. It was then cheaper to carry a spare engine this way but nowadays modern aircraft can fly with one engine inoperative and return to base for repairs so the technique is now rarely used commercially.

While the protracted VC10 negotiations were going on agreement was reached in 1962 between the British and French governments, in association with the British Aerospace Corporation (BAC) and French company Sud Aviation, to build Concorde as a joint venture - the then British Labour Transport Minister, the late Anthony Wedgewood Benn, cuts the tape to allow the roll-out of the first French prototype at Sud Aviation's Toulouse factory. Eight delivery options were placed for Concorde with BOAC hedging its bets and also placing six orders for a US supersonic aircraft, then only a drawing board outline. This began an intensive round of advisory discussions stretching over the next ten years between BOAC and BAC and also in developing relations with Air France on its operational requirements.

The mid 1960s was a very busy time for BOAC engineering. With the VC10 order confirmed work commenced to begin training and establish facilities to maintain the new aircraft. Treforest was extended to allow the aircraft's Conway 540/550 engines to be overhauled there rather than at Rolls Royce in order to save costs. Bringing sub-contract work in-house as a cost saving measure was a recurrent theme throughout the 1950s and 1960s. New hangars were also needed with TBA and TBB being too low to accommodate the VC10s high tail. Fortunately, TBB was shortly to become vacant following the sale of the Britannia fleet and lent itself well to an extension to the roof line at an acceptable additional cost rather than build a completely new hangar.

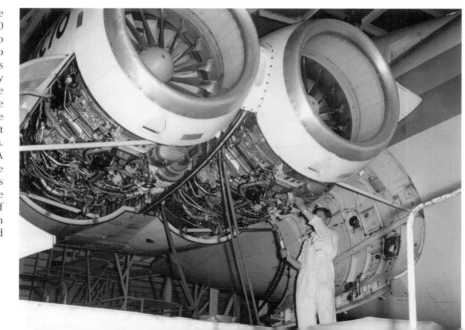

The mid 1960s saw BEA also begin its final fleet modernisation. The Hawker Siddeley Trident and BAC One-Eleven aircraft were introduced and would become their main fleet aircraft for another ten years - the first time except for the very old DC3s and Viscounts that BEA had managed to fly the same aeroplanes for sufficiently long to gain the cost advantages of lower depreciation. Several variants would be introduced over the next five years with aircraft modifications introduced to take advantage of growing technological change, a notable first being the installation of autoland to allow fully automatic landings.

10th June 1965 and BEA's Trident One G-ARPR touches down at Heathrow for the world's first fully automated landing on a commercial flight from Le Bourget, Paris to Heathrow; BEA had worked closely with the manufacturers for several years and been involved in over 600 trial landings in various weather conditions before this first commercial flight. It would take several more years of trials before autoland was allowed on commercial flights in poor weather but is now fitted as a standard feature on most modern commercial aircraft.

Both the Tridents and 1-11s were powered by Rolls Royce Spey engines, the early versions being slightly underpowered but reliable and long-lasting, and quiet, at least on the inside. The Trident was a tri-jet with two engines either side on the rear fuselage and one above with its air intake in front of the fin above the fuselage and removed by a simple arrangement of geared pulleys, as shown here at BEA's Heathrow engineering hangar. It is said that on early flights anxious passengers sometimes enquired if there were any problems with the engines as they were so quiet they thought they had stopped! To give it more power the Trident Three even had a small Rolls Royce RB162 booster engine mounted in its tail to assist take-off, although it was not particularly successful.

With the introduction of the VC10s, BOAC sold its Comet 4 aircraft in 1965 to achieve for the first time a fleet of only two aircraft types and all-jet. BOAC was benefitting from the excellent reliability of the Rolls Royce Conway engines on its VC10s and 707 fleets as well as extended lives on the aircraft themselves. Technology, however, was again moving on and the substantially larger Boeing 747 aircraft was on the drawing board and its prospects and the need to remain competitive meant BOAC had to have them. Options for six were placed in 1966 and planning began for the aircraft's introduction to service. The 747 was so much larger than anything before it that it meant nothing would fit, from airport Terminal piers to hangars, let alone the facilities for handling substantially more passengers on each flight. For BOAC engineering it meant not only a new hangar but just about new everything from towing tugs to access equipment and overhaul facilities for both airframes and engines, the latter also being substantially larger than the existing Rolls Royce Conways. Alongside the introduction of the supersonic Concorde, expected in 1976, itself so new and different everything had to be engineered from scratch, BOAC engineering had a major development task on its hands and looming in the background was the UK government's intention to merge BOAC and BEA into one airline, just three years after the 747 started service.

This page: BOAC's second 747 to be delivered, G-AWNB, outside the new Hangar 01 and being lined up to test it for size. Fortunately, it fitted perfectly. In the foreground an Irish airline, an Aer Lingus 747, undergoes an engine run on the de-tuners. Aer Lingus is now part of the International Airlines Group, as is BOAC's successor British Airways.

Opposite: BOAC's fourth 747, G-AWND, lands at Port Stanley in the Falklands Islands as part of a series of military charters in 1985, fifteen years after its original delivery. It would fly on for another six years and probably have done several more still except it was destroyed in 1991 during the liberation of Kuwait. Such longevity is a common feature of modern aircraft and has changed the face of aircraft engineering and their maintenance practices and procedures.

BOAC's first 747, G-AWNA, about to leave on its final service in the late 1990s. Painted on its port side in its original BOAC livery and British Airways' livery on the starboard side, the aircraft had flown for twenty-six years. Its original maintenance home, Hangar 01, since re-named TBJ, is in the background.

Apart from new engines and the aircraft itself, the most expensive engineering cost for the introduction of the 747 was a new hangar, in fact, eventually, two new hangars. BOAC had recognised that on past experience the growth in the size of new aircraft had been underestimated when designing new maintenance facilities. The new 747 hangars were therefore going to be built in such a way that they could, in future, be extended in three different directions. The roof was to be assembled on the ground and jacked up and the supporting walls built in such a way that they could be breached and extensions subsequently built. The roof could also be raised further if needed. Large sections of the floor were also able to be lowered to allow the removal of undercarriage units and mobile overhead staging fitted so it could be driven into position to allow safe working on the aircraft's tail and fuselage. Started in 1968, Hangar 01 as it was originally called, now designated TBJ, was opened in 1970 and could house two 747s. To cater for its growing 747 fleet requirements, a second hangar, 02, now TBK, was begun in 1971.

Left: The topping out ceremony of Hangar 01 involved the BOAC Managing Director, the late Sir Keith Granville, and the late Hon. Geoffrey Ripon, QC, a Director of Cubitts, the construction company, ascending fifty-four feet to the Hangar roof in a 'cherry picker' to tighten the last nut and bolt. Frankly, I'm not sure I'd have wanted to go up there without a safety harness and a much higher safety rail!

Above, opposite: BEA had also worked out it was going to need more hangar space for its new Trident Three aircraft, expected in 1970, and 'for an Airbus type aircraft' yet to be decided. In fact, the 'Airbus' became the Lockheed TriStar delivered in late 1974 just after BEA had merged with BOAC to become British Airways. The new large hangar, rather unimaginatively designated the 'Servicing Hangar' became nicknamed the 'Cathedral' due to its cavernous interior and high roof. As maintenance needs changed, the Cathedral became more valuable as a so-called 'casualty' unit taking in aircraft overnight for minor repairs before moving back into service the next day.

Trident three
British airways

A combined effort to give you the best airline in the world.

British Airways

3

ENGINEERING A MERGER

1 April 1974 is a key date in British Airways' engineering history, not least because it was the date when BOAC and BEA were merged to form today's British Airways. At a stroke, their two separate engineering departments were joined together, a combined workforce of just over 14,000 people that also included a separate engine overhaul business and four small regional airline subsidiaries plus British Airways Helicopters. Despite what would appear to be a good opportunity to reduce numbers and eliminate duplication, within a year the total engineering staff employed had actually grown to nearly 15,000, approximately twenty-eight per cent of British Airways' workforce; the group comprised a wide range of skills and different work practices that was so extensive it required nine separate trades unions and five national sectional panels (NSPs) to negotiate wages and conditions of service.

In fact, there was no immediate, physical merger except in name. BOAC and BEA both operated and engineered aircraft but any other similarities in what they did were not as easily assimilated as some may have thought at the time. What was immediately apparent, however, was the need to

Opposite: This is one of British Airways' early advertising posters to promote the new company as a combination of the best of its airlines' skills and services, including engineering. It may have become the best airline in the world but the amalgamation of such a wide range of different aircraft types and working practices would be a very difficult task for the new company's management to sort out over the following years.

ensure that the previously separate companies continued to operate smoothly and present a united face to the world and its customers. The name 'British Airways' became, in effect, merely a cloak to disguise the fact that BOAC and BEA continued to operate almost exactly as they had done before the merger, but in a common livery and under a common brand name. The two old companies just became separate parts of the new company organisationally identified as 'British Airways Overseas Division' and 'British Airways European Division'. Just to complicate matters further, a third set-up, called 'British Airways Regional Division', swept together the UK domestic and regional operations that BEA had previously also carried out. Within these three divisions of the new British Airways, everyone guarded their territory and looked to achieve an organisational advantage over the others. What had been established for common sense, practical reasons in fact became a block to the gradual merger of the three divisions, not least in engineering.

In late 1974, just a few months after the merger, a glossy and impressive engineering brochure was issued under the title of 'British Airways: European Division Engineering'. It was a sales pitch to advertise the Division's engineering facilities to third party airlines that were prospective customers, and to all intents and purposes it appeared to be from British Airways itself. There was no reference to the Overseas Division and the images in the brochure very carefully showed only

BEA's old engineering facilities, despite Overseas Division's being a stone's throw away. This was a good example of the old BEA attempting to exert itself and take a premier position in the new British Airways' engineering future; it certainly did not take forward the imperative need to assimilate the two engineering operations quickly and efficiently at a time when world aviation was rapidly being opened up to new airline competitors eagerly snapping at the heels of the old established, often nationalised, so-called 'legacy' airlines, invariably with much higher costs, not least operational and engineering costs.

It took eight years of long drawn-out discussions to begin to unify and reduce staff numbers forced by economic circumstances that, without the UK government's financial support in the 1981/2 financial year, would have bankrupted the airline. Without this impetus it is questionable how long it would have taken for management and the trades unions to reach an agreement to re-organise in any meaningful way in order to improve efficiency and reduce costs. Many staff, even in management, were against the merger in the first place, believing it a wrong decision by the government to unite two quite different organisations and prospectively weaken both in the process. There was also little incentive to re-organise into an optimum structure with the threat of disruption and loss of jobs being the likely result. It was inevitably going to be a long drawn-out process to get anywhere near the preferred size and shape of a new engineering organisation let alone the new airline as a whole, and so it proved.

1983 would see the final size and shape of British Airways' new engineering organisation begin to emerge. A key management decision was taken to bring together all operational aspects of the airline so that engineering and flying, i.e flight crew and cabin crew, as well as ground handling, catering and logistics came under one director reporting directly to the then new Chief Executive, Colin Marshall, later Lord Marshall of Knightsbridge. This made the new operations department the largest employer of labour in British Airways – 24,000 staff – with a budget of £1.4 billion, with engineering's primary objective to be supportive of the operation as a whole and a significant contributor to improving operational performance, not least punctuality. This brought a very clear focus on customer service. Colin Marshall had made it very clear to his executive team that the airline only existed to serve its customers, some had previously thought it was the other way around. Unless operational performance could be improved British Airways would not be able to meet its new corporate objective 'to be the best airline in the world'.

Engineering's contribution could be summarised, simply, as the delivery of safe, reliable aircraft on time. It was, of course, much more complicated than this implies. It was a three-dimensional challenge: a combination of continuing to satisfy the essential technical regulations that govern the operation and maintenance of flying commercial aircraft, building an effective new organisational structure to deliver it and continuous improvement to maintain it. It was a very difficult three-dimensional challenge too, given that not only had significant changes still to be made to blend together its two predecessor engineering organisations, but at the same time to complete an essential fleet rationalisation that had begun in the later 1970s. This was based not only on selling or scrapping the older, maintenance-heavy aircraft such as the 1960s vintage VC10s, 707s and Tridents, but to introduce new more efficient aircraft into the fleet, which had begun in 1980 with the first deliveries of the new Boeing Super 737.

As the older aircraft left service so did an increasing number of their old engineers, including my father. Then in his early 60s he and a number of other old-timers had been re-finding their 'make-do-and mend' skills in the reclamation unit in Technical Block 'B', or 'Wing hangar' as it was known, where Concorde was maintained. The reclamation unit appeared to be, at

One of the first of the new British Airways Engineering's tasks would be to repaint its aircraft into its new livery. A quick fix was to replace the name with 'British Airways' on either side of each aircraft's fuselage but it would take several years for the whole fleet to be completed; the 747s were not even started before late 1975. These Tridents in BEA's West Base hanger have all been re-named while BEA's old fuselage paint colours and tailfin logo remain in place. At the back of the line there is even an old turbo-prop Viscount similarly renamed.

least to me, a sort of pre-retirement occupation using traditional skills to fix all manner of bent bits of aluminium aeroplane components rather than scrap them and buy new ones, an admirable approach but there was a growing realisation that sometimes new ones cost less than the cost of repairing the old ones. Another realisation was that a growing number of its older hangers, like TBB, were becoming not fit for purpose as larger modern aircraft came into British Airways' fleet. The new aircraft not only did not fit the old hangars very well but they also did not need to be under cover quite so much given the periods between major maintenance checks were getting longer. Although it would take two more decades to really take on a new shape, the old BOAC and BEA engineering bases gradually began to be transformed into a homogenous area as hangars were dismantled and new aircraft parking areas were created. The completion of what had been a difficult amalgamation of the BOAC and BEA engineering organisations opened up the opportunity to demolish many old hangars so that in the 21st century a very large aircraft parking lot now covers most of what was BEA's old engineering area.

Left and opposite: A double-page spread from British Airways (European Division) engineering brochure advertising its skills and services. When the original air-to-ground photograph was taken it also showed BOAC's old engineering base immediately to the east but was carefully doctored to omit it for the brochure. In effect, it was as if BOAC engineering did not exist and any other airlines looking for maintenance services should apply to what had been BEA. In fact, the BOAC engineering facilities were huge with three major hangar and workshop complexes, as the second image shows; it is unusual to see so many parked aircraft in the engineering base which suggests there may have been an industrial dispute so the aircraft had to be parked somewhere.

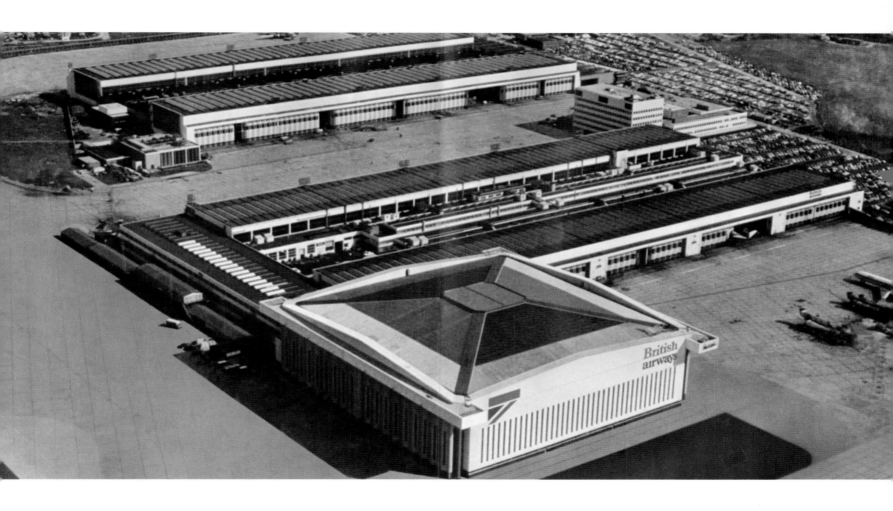

Can We Help You?

At your request a British Airways consultant will be pleased to discuss your problem with you and advise how British Airways can be of assistance.

With the backing of such extensive British Airways support facilities and management expertise and experience you can be assured of a totally professional, practical and objective analysis of your problems.

Opposite: A few years later British Airways Engineering brought out a new brochure, this time better balanced but still with a front cover image of the old BEA cathedral hangar and with a Tristar aircraft undergoing maintenance, although now in the new British Airways' livery. To be fair, BEA had expertise in TriStar maintenance being one of the first airlines to offer Tristar third-party work; BEA already had a major contract with Gulf Air for their TriStars and with third-party contract work a major prize for British Airways, it made sense to maximise promotion of one of its prime assets.

Right and overleaf: By 1978 British Airways was going all out for third-party work across all its fields of expertise and not just engineering. With many surplus staff across all its Divisions it was an opportunity to bring in extra work to improve its manpower performance and also increase revenue, an example being BEA's significant experience in a range of helicopter operations, from scheduled services to charters serving the North Sea oil fields.

Opposite and next spread left: When British Airways was formed its combined fleet of aircraft comprised twelve different aircraft types including helicopters and nineteen different aircraft engines, a significant and complex organisational task to grapple with. Into this mix in 1976 came Concorde, not just another jet but a supersonic one with all the extra complexities and new technology that came with it. Originally ordered by BOAC and expected to be in their livery when launched, the merger had overtaken that and this artist's impression of Concorde on the ramp at Heathrow backed by BOAC's Super VC10s never actually materialised. A British and French joint engineering triumph became a triumph for British Airways Engineering that began on Concorde's maiden commercial flight to Bahrein on 21st January 1976.

Next spread right: British Airways first Boeing Super 737 under construction at Boeing's Seattle plant in 1979. Registered as G-BGDC and delivered on 8 February 1980, the 737 would become the backbone of British Airways' short-haul operation for the next thirty years. It was at the time the most technologically advanced airliner of its size and type and would take over the operations of the old BEA Tridents that were rapidly becoming obsolete and expensive to operate. It was considered an advanced type because of a range of avionics and noise supressing modifications that would improve the aircraft's performance and fuel economy by up to 20% over its predecessors. The development of avionics was rapidly growing in the early 1980s and would become an increasingly important part of British Airways' future engineering profile and expertise – a good example being a major modification to the 737s autoland system that Boeing were unable to complete before planned delivery and which British Airways Engineering successfully installed and below cost. (Courtesy of Boeing)

British Airways urgent need in the early 1980s to begin a programme of fleet simplification had been reinforced by a continuing rise in fuel prices and increasingly strict noise regulations. The future fleet would be based on four different subsonic types with the 747 leading for long-haul services, supported by the TriStar, and the new 737 and older BAC 1-11 operating on European and UK domestic routes. They would shortly to be joined by the new Boeing 757 in January 1983 which would progressively take over many of the 1-11 routes. For engineering this would lead to major changes with longer maintenance and component lives, especially on the new aircraft, progressively minimising workloads. The very reliable and fuel efficient Rolls Royce RB211 family of engines that powered the 747s and TriStars, would also power the 757. British Airways' first two 757s being built at Boeing's Seattle plant already have the beginnings of its livery on their tail fins. When fully painted, just the word 'British' would be added on the fuselage as a short-lived attempt by the airline in the early 1980s to project its brand name more strongly. G-BIKA and G-BIKB were the first two deliveries named as Dover Castle and Windsor Castle respectively. (Above, courtesy of Boeing)

Some real aeroplanes combined with computer generated graphics were the stars of the British Airways' 2011 advertising campaign that highlighted its 'flying know-how', although the contribution of its engineers over the decades became a casualty of the cutting room floor. Some actors had been dressed as 'engineers' (although they were in entirely the wrong outfits), but were not shown in the final film. (Bartle Bogle Hegarty)

4

ENGINEERING AN AIRLINE

Safety is a core business value that underpins everything that British Airways does; key to that value is how the company operates and engineers its aircraft, a pairing of flying skill and technical know-how both in the air and on the ground. 'Flying Know-How' has an advertiser's ring to it and, indeed, is where it came from. In 2011 British Airways' advertisers came up with a campaign that majored on its heritage credentials and operational skills, highlighting almost one hundred years of flying experience. This was flying experience in its broadest sense bringing together a very wide range of skills and deep-seated technical knowledge about operating a very large international airline across the world. The campaign was a dip into nostalgia that left many in the business misty eyed. Lots of old aircraft flying around, including a 1920s biplane and Concorde - actually, the ninety year old biplane did fly but Concorde was all old film and clever graphics - flight crew resplendent and dashing in their uniforms, yes, some of them were real pilots, and actors making a very good job of pretending to be excited passengers about to depart. What was notably missing, however, was any visible reference to the substantial contribution of the engineers to that 'flying know-how' so graphically represented.

I suspect nobody thinks of that today as they wait in Terminal 5 at Heathrow admiring the view or the duty-free shops. If thought about at all it is probably considered of little consequence. Provided flights operate to schedule there is an expectation of their safe and punctual arrival. Thirty years ago a passenger survey discovered exactly that with British Airways' professional technical integrity and standards of excellence were taken for granted. In one sense that view is a satisfactory compliment, but within British Airways nothing is taken for granted and the 'operation' is a constant focus of attention. For the operations department, including engineering, satisfying the customer is everything and engineering's principal customer is British Airways itself. The story goes that Colin Marshall, following his appointment as Chief Executive in 1983, had installed in his office a small videocom terminal that showed the arrivals and departure times of all British Airways' flights. If one was late the Terminal Controller would soon get a 'phone call to ask why. It certainly placed a focus on improving aircraft punctuality which Colin Marshall knew was a major part of the bedrock of the airline's reputation.

1983 was a turning point in another significant way because of the Conservative government's commitment to privatise British Airways, targeted for the later 1980s, but several years of consistent and substantial profits would first be needed to demonstrate to prospective investors that the company would be an attractive shareholding. From an engineering perspective, this brought a very close focus on costs and how these could be improved without compromising safety.

Engineering costs represented ten per cent of the airline's operating expenditure at the time with manpower costs a third of that. Costs had already been reduced quite dramatically during 1981/82 - manpower costs coming down nearly thirty per cent across the airline with engineering staff numbers down to 7,900 by 1983. Re-shaping this core complement would be a key element in bringing costs down further and was facilitated by a review of engineering procedures and processes across the organisation. Many procedures were based on traditional methods that had not caught up with improving aircraft, engine and component reliability and longer maintenance periods, particularly on the newer aircraft coming into the fleet. For example, by changing maintenance schedules to concentrate the majority of the work in the winter months, when operational demand for serviceable aircraft was lowest, it freed up not only more aircraft for the summer peak but allowed the workforce to be re-directed in the summer months to servicing third-party work, a growing source of profitable business, and to carry out any major modification work on British Airways own aircraft. This refocusing on better planning and maintenance scheduling also had the effect of improving the delivery of aircraft on time for service, a notable example being the 747 major maintenance line, '...which had run late for as long as anyone could remember'. There was also a review of what engineering work was actually being done. For years the policy had been to engineer everything in-house where it was technically feasible to do so, almost regardless of cost. The review highlighted the opportunity to out-source the resource-draining, low technology work to other companies who were better able to supply it more cheaply and released manpower to take on more lucrative third-party work. My father's old reclamation unit rapidly disappeared.

Third-party work was seen almost as a panacea to improving engineering's cost performance. BEA had led the way in the early 1970s with a major contract with Gulfair to maintain their Lockheed Tristars. Prior to that neither BEA or BOAC had carried out any third-party work of any significance other than engine work undertaken by BOAC's subsidiary engine overhaul company at Treforest in South Wales. In the early days of British Airways third-party work also continued to be concentrated on Treforest with the engine business rapidly growing to become one of the largest engine overhaul facilities in the world; the Queen's Award for export was awarded to the company in 1980. With a focus now on the potential for third-party work, particularly for the growing number of smaller new-entrant airlines who would struggle to build their own maintenance organisations, the opportunities looked bright. The business was certainly there in the 1980s and British Airways had a pool of talented, highly experienced maintenance engineers and spare capacity. It was an obvious business decision to make and the business began to grow quickly, from the simplest washing of aircraft to 747 major maintenance checks. British Airways Engineering began to build an enviable reputation for quality and workmanship across all maintenance fields including mechanical and avionic work and even such specialist areas as flammability testing of aircraft interiors and non-destructive testing (NDT). The business was becoming almost a brand in its own right and while third-party work had originally been seen as an opportunity to undertake other airlines' engineering during gaps in the maintenance line for British Airways' own aircraft, there was a growing perception that much more could be done. The balance, however, was to ensure that the airline's own requirements were not compromised by the pursuit of third-party profit in its own right. British Airways' main purpose, after all, was to

A view from Terminal 5 on the upper level just to the left of the W H Smith bookshop – browsers may glance at the scene but the major contribution of British Airways' engineers in ensuring the aircraft in view are on their stands on time and ready to go is invariably overlooked.

sell its commercial services and prioritise its engineering operation to support that objective.

The next few years would see almost continual change affecting British Airways, particularly in improving its overall profitability and brand offerings as it headed for privatisation in early 1987. Many of these changes affected engineering and at a time when improving costs was a continual background theme. A change in British Airways' livery in 1985 was a major task to accomplish with over 250 aircraft in the fleet needing repainting over a short time frame. Even the engineers themselves were to be re-dressed, for the first time being put in new uniforms, they were too smart to be called overalls, as an acknowledgment that they had a public face and needed to be suitably dressed for the part.

Appearance, performance and reliability continued to take centre stage with a major push on improving aircraft appearance both outside and inside the cabin. The introduction of seat-back in-flight entertainment and other electronic gadgetry all needed to be kept working all the time and not just left as a defect to be fixed at the aircraft's next check. There was a rapidly growing acceptance that British Airways Engineering's end product was no longer just a piece of high quality engineering but also a satisfied passenger. It was the catalyst to begin the move away from being a high cost factory to an essential and integral part of the challenge in delivering the British Airways' brand.

Opposite: TBA, South Pen hangar, October 1984. A 737, BAC 1-11 and Trident Three (from top to bottom) undergo a light maintenance check before returning to service. In fact, the 1-11 looks like it is undergoing a major check including a wing modification and engine change as well as interior work. This may be one of the early examples of major modification work being re-planned for the short-haul fleet to be undertaken during the winter months.

As the decade progressed further challenges arose. In 1988, the integration of the British Caledonian Airways' aircraft fleet following BCAL's takeover by British Airways put a new dimension into the mix. As well as two new aircraft types, the Douglas DC10 and Airbus A320, the first Airbus type for British Airways, BCAL's maintenance was ninety-five per cent sub-contracted and not worked in-house. To save on sub-contract costs and maintain its productivity levels, much of this work would be brought into British Airways' workshops at Heathrow with the engine work going to Treforest, at the time experiencing its own extension and modernisation close by at Nantgarw. This eased the integration from a cost perspective but it was an additional complication when the business had its work cut out to re-shape itself for the 1990s.

Re-shaping was an absolute necessity if engineering was to meet British Airways' expectations for a growing fleet, at least forty new aircraft by 1995 worth US$7billion, when engineering's resources at Heathrow and Gatwick were close to capacity, particularly for 747 heavy maintenance. Devolution became the term to describe the process of searching for additional heavy maintenance facilities as well as for avionics and interior work. Space and land at Heathrow was very expensive and there was a shortage of skilled engineers. Various overseas options were considered over the next few years but all were found wanting. South Wales became the favoured location, not least because of its siting close to the modernised British Airways' engine facility at Nantgarw and a skilled local workforce. Importantly, there were also very generous Welsh Development Agency grants available.

The first Gulf War in 1991 changed many things not least the expectations and plans of the commercial aviation industry which needed to react quickly to bring back lost business as world trade and travel adjusted to a major economic downturn. Saving costs and improving productivity while not compromising on safety or standards had never left British Airways Engineering's agenda, but the war gave it further impetus. Nantgarw, or to give the subsidiary company its official title, British Airways Engine Overhaul Limited, was sold in December 1991 to General Electric Aircraft Engines, a US company whose GE90 engines would subsequently power British Airways new 777 aircraft expected in 1995. GE bought not only the Nantgarw engine works and its substantial third-party engine overhaul contracts (although not the engines themselves), but also took on all British Airways' engine overhaul requirements previously undertaken both there and in TBD at Heathrow, a contract worth £100m + on engine overhaul and related services; this was a major move away from engineering its own aircraft engines and was the first substantial shift towards outsourcing any of its engineering requirements. More would follow as the 1990s trend towards globalisation began to develop as did world economies, not least in places such as Asia and the Far East where specialist aircraft engineering maintenance facilities began to expand and compete with the traditional suppliers, the legacy airlines themselves such as British Airways.

Globalisation began to be seen as something that British Airways needed to respond to; if it was to remain a global airline it would "need support from a successful global engineering business". The seeds were being sown to move British Airways Engineering away from the burden of being a cost centre and towards becoming a profit centre business in its own right with third-party work leading the way. Moves to turn the engineering business into a limited company began in 1996. It was feared it would be the precursor to another subsidiary sell-off like BAEOL, but management were more concerned with making the business "..more commercially aware, provide it with greater freedom to act independently and offer better value to customers", the main one being British Airways. At the time British Airways provided eighty per cent of engineering's workload with twenty per cent being third-party. The aim was to build third-party work to balance the split at fifty-fifty, contracts already being in place with over one hundred airlines. Even the US domestic airline maintenance market was being chased as part of a joint engineering venture with USAir. At the time British Airways had a substantial share-holding investment in USAir so the joint-venture made sense, but it eventually fell away as the investment was sold and links were established with American Airlines that exist to this day.

Business Process Re-Engineering (BPR) was the title given to the first steps in moving towards a limited company. Expected to boost maintenance throughput by between twenty and fifty per cent it was aimed at ridding the department of issues that at the time were believed to impair productivity and profitability. In other words, engineering was to be streamlined by reducing staff numbers over the following three years, through natural wastage and early retirements, and become 'a leaner, fitter and more competitive workforce...'. Going beyond the management speak, the ultimate aim was to become less

Opposite: The interior of 1-11 G-BBMG undergoing a major maintenance check that was required to be undertaken every 7000 flying hours or four years, whichever came first. The check would take about 10,000 man hours of work to complete and were worked to an old-style maintenance schedule that required the removal of components based on the number of hours flown rather than on their actual condition, an expensive way to maintain aircraft and with no noticeable improvement in safety standards. The interior has been stripped completely so the fuselage structure can be properly examined and, where necessary, non-destructive testing (NDT) carried out.

ONE-ELEVEN MAJOR

reliant on British Airways' work and secure a larger percentage of the world's engineering business, an aim with both risks to the parent company and opportunities for the subsidiary.

Within only eighteen months the focus had changed entirely. BPR continued apace but the focus for engineering in future years would unquestionably be on British Airways' own requirements. Third-party work would still be carried out where capacity existed but on a selective basis and only where it was clearly profitable to do so. What had become clear and had precipitated the change, was that the world airline engineering market was not developing as widely or as quickly as some had forecast. Lower demand had led to a worldwide over-supply of maintenance capacity with very soft prices and over-enthusiastic selling leading to some contracts becoming unprofitable. The perceived opportunities had themselves become serious risks and the pursuit of subsidiary profit was permanently shelved.

A worldwide over-supply of maintenance capacity was the least of British Airways Engineering's problems at the turn of the century. The grounding of Concorde in 2000 following the unexplained loss of an Air France Concorde at Paris was bad enough, but its effect paled into insignificance compared to the changes that had to be made across British Airways as a whole following the 9 September 2001 (9/11) terrorist attack in New York. The 2001/2002 financial year was the worst in British Airways' history. Profits declined by seventy per cent over six months with the long-haul business travel market particularly badly hit; the European market also faced growing and intensive competition from the so-called European 'no-frills' airlines such as easyjet and Ryanair. It was another defining moment across the company as a programme called 'Future Size and Shape' was put into effect to fight back. Staff would take the biggest hit with a twenty per cent reduction across the company aiming to reduce staff numbers by over 13,000. Aircraft types were to be reduced, mainly some of the smaller ATPs and 737s, a total of thirty nine aircraft in the first instance, but all contributing to a projected cost saving of over £850m. In 2003, even Concorde would retire as its business passengers declined to an uneconomic level. Gatwick operations were radically changed, moving from being a 'hub' operation to point-to-point services i.e direct services, with consequential changes in aircraft types to fly those routes. Just to give everyone in the airline an additional focus, a target of achieving a ten per cent operating margin was also thrown into the cost reduction mix.

British Airways Engineering would of necessity be at the forefront in achieving that target while still being pressured to maintain and engineer the company's aircraft without compromising on safety standards and to deliver them on time to service. 9/11 actually created an opportunity for British Airways to start thinking radically about how it engineered its aircraft and to consider what in previous years would have been considered unthinkable. The airline had for many years customised its aircraft and equipment and even its maintenance schedules, but this often came at a high cost. By adopting industry standard configurations and modifications rather than bespoke ones, it would be possible not only to save on the initial costs of customisation but also on the costs of reconfiguration when the aircraft were eventually sold on; most prospective customer airlines operated to industry standards and did not want a British Airways hand-me-down. British Airways had learnt some expensive lessons when it had sold on its highly modified 747 -100 and 757 aircraft. It could not afford to do that again, especially as aircraft were now being sold on or handed back to lessors much earlier than before in order to keep fleets young

A British Airways 747, G-AWNO, prepares to leave the TBK paint bay following its re-paint into the new British Airways livery that was launched in December 1984. G-AWNO was the first aircraft to be re-painted. The large pile of brown paper and plastic that is being cleared away of the hangar floor is the masking material used during the aircraft's spraying. Spraying the wings and fuselage were all done by so-called 'airless' hand-operated spray guns but there was still some spray mist that required a lot of masking material. The British Airways 'coat of arms' design applied to the tailfin was a more delicate hand stencilled job. It would take 933 litres of paint for a 747 (about enough for ninety-three cars in those days) and take around twenty days to complete, including removing all the old paint with paint stripper scraped off by hand.

Above left: British Airways new livery also included a new interior on all its aircraft, another massive refurbishment project to be completed as soon as possible. British Airways Engineering was highly skilled in interior modifications having for many years always fitted its own interior designs and specifications in its aircraft, a costly process but one which was considered to be important in building and maintaining British Airways' brand value. These engineers are completing the re-fit on a 757.

Below left: With so many aircraft to repaint and fit new interiors, British Airways Engineering just did not have enough staff to carry out the work within a reasonable time. It was decided to engage Marshall of Cambridge (Engineering) to undertake the work on two long-haul Tristars, both pictured here in Marshall's Hanger 17 at Cambridge airport. It was also a bit of a reunion for the aircraft in that two other ex-British Airways -500 Tristar aircraft were also being worked by Marshalls for the MoD having been sold to them to convert into air-to-air refuelling tankers.

Opposite: British Airways new livery included for the first time a standard 'uniform' for engineers. While engineering overalls and suitable outside workwear had been standard for many decades, they were invariably almost indistinguishable from any other company clothing apart from differences in each company's name. The new livery by the US design agency Landor changed all that. In future, if engineers were seen at all they would clearly be identified as part of British Airways and be suitably dressed for the part.

and retain the benefits of increased efficiencies that new aircraft and engines offered.

With the very serious financial effects of 9/11 very clear in everyone's minds it was also an opportunity to make some major changes to out-of-date working practices and engineering shift patterns. Management and staff from the European short haul fleet operation, who had faced the most serious and growing competitive threat from the 'no-frills' airlines, introduced significant changes to fleet plans and schedules to provide more aircraft availability during the day when demand was at its highest. 'Fly by day – fix by night' became the catchphrase, not unlike BEA's way of working decades earlier although this time to 21st-century standards. A significant and major part of those new standards was the introduction of a fully integrated computer system across British Airways Engineering known as 'EWS' (Engineering Wide Systems). This was a huge and complex task but the goal of removing incompatible 'legacy' systems and creating a common means of communication and integration across British Airways' operation and with its partners, suppliers and overseas stations, had to be grasped.

It would take several years for the benefits of EWS and maintenance simplification practices to work though, but by 2006 British Airways' fortunes had begun to improve. The ten per cent operating margin was reached in 2007 and the aircraft fleets began to be upgraded, especially at Gatwick with the Airbus A319 replacing the older 737s and firm orders being placed for additional Airbus A320 and A321 aircraft. For long haul routes, the first orders for the huge new Airbus A380 aircraft and Boeing 787 were also placed, two very advanced aircraft that would introduce a significant change in aircraft maintenance in that they would be the first ever commercial aircraft to be 'e-enabled' i.e having

the ability to provide automatic in-flight monitoring of systems and performance. These aircraft would continue the trend that each new generation of aircraft tended to introduce reliability improvements and technological advances that would reduce the maintenance load on their engineers. This was especially so with the A380 and 787 which would bring not only opportunities but challenges to work the aircraft in fundamentally different ways than before to obtain the full benefits that their new technologies offered.

Embracing and maximising the benefits of new technology would become one of several strategic objectives of British Airways Engineering's 2011 plan to 'Engineer our Future'. Before that, however, it had to weather yet another downturn in the fortunes of the aviation industry as it was very badly hit by the banking crisis of 2008. While being well prepared and well-funded, British Airways could not escape and had to introduce further significant cost reductions across the company with engineering staff numbers reduced to 5,000, itself a reflection of the falling numbers of engineers now needed to manage a total fleet size of 238 aircraft. During the crisis, thirty-five airlines went bankrupt or were forced into rushed mergers. For its own part, British Airways announced that it was proposing to merge with Iberia Airlines of Spain and become part of a new airline consortium, The International Airlines Group (IAG). IAG was not another global airline alliance such as oneworld, of which British Airways and Iberia remained members, but would bring the two airlines together under one owner who could maximise the synergies such a merger would bring, not least in aircraft procurement and engineering. The merger was completed in November 2010 and within two months British Airways Engineering together with Iberia Maintenance and AJ Walter Engineering had already secured a major five-year, multi-million deal to maintain Thomas Cook's A320 and A330 aircraft and was followed by a second contract to maintain the wheels and brakes on Cook's 757s. The Thomas Cook contract was an example of British Airways Engineering returning to the third party maintenance market with a bang. There had been job losses due to the banking crisis but there remained surplus manpower that could be usefully used to tap back into what was still a lucrative third party market, provided it was carefully costed, but this time there would be no mistaking that British Airways Engineering's requirements came first.

By 2011 British Airways Engineering had weathered the downturn well and was preparing to introduce the 787. It had originally been expected that the A380 and 787 would be introduced a year apart, but due to delays on the Boeing production line both aircraft were now to be delivered almost a few days apart in 2013. Successfully meeting the challenge of any new aircraft introduction is a major accomplishment but two together was unprecedented in modern times, not least as they were the first truly new aircraft introductions for seventeen years. This new milestone in the history of British Airways Engineering was very much based on innovative ways of working to make the most of the new aircrafts' technological capabilities and expectation of reduced time on the ground. In effect, engineering an airline in the 21st century would now begin in the air rather than on arrival at its destination, truly a step change away from the previous century when British Airways began nearly one hundred years ago.

Left: Royal Flights often required a special interior fit-out. When Her Majesty flew to Portugal in 1985 the Tristar aircraft used was truly fit for a Queen with a gold (coloured) carpet and special divans, lounge seats, coffee tables and wardrobes. The Royal Mod (Modification) Kit as it was known had been used on other Royal flights but needed some special modifications to fit the Tristar, including attaching to the forward bulkhead the silver St. Christopher medallion that always travels on all Her Majesty's flights.

Opposite: One of the first third-party contracts following British Airways Engineering's drive to secure additional revenue paying work during the summer months, was Gulf Air's 747 Combi aircraft N203AE seen here in TBJ on 31 March 1985. The Combi was a part passenger/part cargo 747 and was to undergo a package of 'special inspections and checks, modifications and defect rectifications'. The word 'special' is important in that while British Airways engineers were very good at both standard and unusual third-party work, it needed very careful planning and procurement skills to ensure that the work was carefully priced and profitable.

Opposite: To scale up the challenges presented by an even larger aircraft undergoing a complete cabin replacement as well as a major maintenance check at the same time, this interior image of British Airways 747 G-AWNC in TBJ rather sums it up. Apart from the interior work, major structural changes involved replacing several floor beams, engine pylon and undercarriage reworking and replacing sections of the aircraft's lower fuselage skin. It would take 250 men per day a total of thirty nine days to complete the works, approximately 63,000 man hours, and required some new thinking to plan and co-ordinate the many and varied tasks involved.

This page: Apart from the major task of repainting all its fleet aircraft and refitting their interiors, 1986 saw British Airways Engineering embark on the biggest engine modification programme by any airline worldwide. The three main aircraft engine manufacturers, Rolls Royce, Pratt & Whitney and General Electric, constantly vied with each other to produce engine improvements, particularly in terms of engine thrust and fuel consumption. British Airways 747-236 aircraft were by then nearly ten years old and powered by the early Rolls Royce RB211-524B engines. Although they had had several modifications they were becoming uncompetitive, but Rolls Royce had developed a modification programme, the 'D4 upgrade', to bring them to the latest specification. The modification was a huge undertaking involving almost a complete rebuild of sixty engines plus spares and costing £100m. In effect, for the (then) cost of one 747, all British Airways 747s would be compatible with the latest models and allow ultra-long sectors to be flown non-stop, such as London-Tokyo.

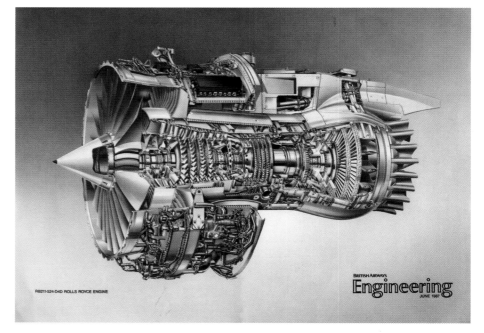

RB211-S24-D4D ROLLS ROYCE ENGINE

This spread, overleaf: 1986 is a good year to review the first ten years of Concorde's engineering operation. Concorde was a thoroughbred in the true sense of the word in that it was special and needed special care to get the best out of it. Its high costs were a particular challenge with maintenance man-hour costs around two and a half times that of a 747 and spare parts with very inflated price tags – an engine cost about £2m. Despite being a thoroughbred it was actually 'old technology' being a 1960s design with lots of hydraulic systems; these leaked a lot in the early days until the fault was traced to hardened seals needing tens of thousands of them to be changed across the fleet of seven aircraft. Being special it had a specialist team of around 200 engineers that looked after the aircraft throughout its service, a unique feature in modern aircraft engineering but which paid off in the level of expertise and co-operation that was built up in the team and between them and the operational flight and ground crews.

Just past its tenth anniversary, British Airways' first Concorde, G-BOAA, had clocked up 12,000 flying hours during 4,150 flights with nearly 4,000 braked landings and had travelled approximately 12 million miles. It was time for what the BBC at the time rather simply described as an M.O.T. Unlike one's car, however, this would be a major maintenance check and the first for any Concorde involving a minute inspection of the aircraft's whole structure. All the engines and associated moving parts were removed including the electrical wiring and looming and several modifications were carried out including work to extend the aircraft's service life well into the next decade. A final repaint and a brand new interior plus a polish would complete the works. What was surprising was that the aircraft was in very good condition with virtually no corrosion, due to the very high operational temperatures reached of up to 127C preventing moisture deposits. This image was used on the front page of a British Airways Engineering magazine and was, according to management, put in as a subtle safety training exercise to see how many engineers spotted the many safety breaches displayed in the image.

Next spread, left: By 1987 third-party work was rapidly growing. From Trans World Airlines 747's to KLM helicopters and just about every combination of work in-between was coming into British Airways' hangars. This work was supposed to fill in the time when there was spare capacity in British Airways own maintenance lines, but there were instances of the allocation of specific engineering resources to chase what was seen as a significant source of revenue. TBA's hangars were beginning to be occupied with other airlines' aircraft, such as this Aer Lingus 1-11 undergoing a 'D' check and a CP Air 737 a '4C' check in the TBA South Pen.

Next spread right and overleaf: British Airways (and its predecessor BEA) had operated out of Birmingham airport since the Second World War. There was an engineering organisation but not a lot else as aircraft were maintained in one leased bay of an old 1930s hangar where even the doors often failed to operate properly. All that changed in 1988 with a major investment by the airport authorities to renovate and develop the hangar facilities to modern standards. Resplendent in British Airways new livery, a 1-11 is pushed back into the new facilities. The 1-11 was the mainstay of British Airways' operation at Birmingham which was an important regional hub for the company, an importance reflected in the line-up of 1-11s on the ramp area of the passenger Terminal ready to go out on service.

British Airways Gatwick based charter airline, British Airtours, also had its own engineering operation but it existed alongside and separately from British Airways Engineering's own Gatwick team. It would not escape the scrutiny of engineering's drive for efficiency and lower costs and the perfect opportunity presented itself to merge the two organisations when British Airways took over British Caledonian Airways in 1988. Although BCAL had a large Gatwick-based engineering organisation and hangar complex, a considerable amount of its engineering maintenance work was outsourced, the cost of doing that could be lowered by bringing it back in-house and merging all three engineering operations into one.

This spread: British Airways' takeover of British Caledonian Airways in 1988 introduced both the Douglas DC10 and Airbus A320 into the airline's fleet. British Airways Engineering had no expertise in either aircraft but BCAL was considered one of the leading European airlines on the DC10 with eleven DC10-30s including three from its charter subsidiary, Cal Air; it had also been undertaking a considerable amount of preparatory work and training for the introduction of the A320. With differences in agreements and practices, standards and procedures let alone pay and conditions, it would take some time to merge successfully the three organisations (including British Airtours Engineering) with one of the first steps to bring back the outsourced DC10 heavy maintenance line and to recruit additional staff to take on the additional work load.

DC10
MAJOR

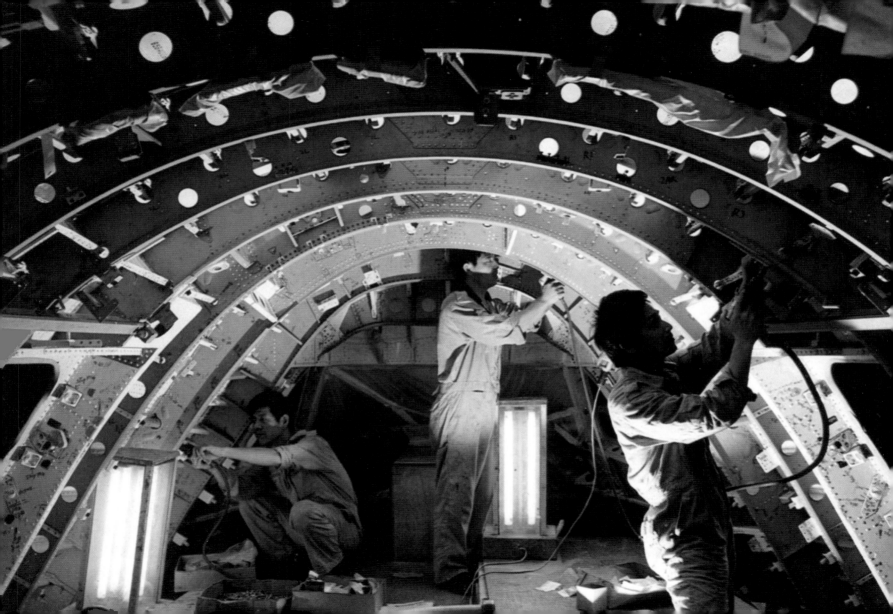

Opposite: Modifications to existing aircraft have always been a major part of aircraft engineering maintenance and none more so than the so-called Section 41 modifications carried out between 1990 and 1994 to extend the lives of British Airways' old 747-100 aircraft. They had been due to retire when the new long-range 747-400 aircraft were delivered but with a planned expansion of services more aircraft would be needed and extending lives rather than buying new was cheaper. The modification programme was extensive, however, and British Airways Engineering just did not have the resources in both manpower and hangar space to carry out the works. HAECO of Hong Kong were awarded the contract which involved, amongst many other things, replacing all the fuselage frames in the nose and forward part of the aircraft (the Section 41 area hence the name) and the massive straps which hold the fuselage on to the wing. The 'Phoenix project' as it was appropriately called, would take up to 140 days to complete each aircraft. It was more than a life extension programme and more like re-manufacturing each aircraft.

Right: British Airways Engineering's 747 maintenance work was undertaken in TBK and TBJ, the old 01 and 02 hangars it had taken over from BOAC in 1974. Nothing else was big enough and the hangars were often filled with aircraft undergoing work as part of the 747 maintenance cycle. The cycle is based strictly on flying hours with a daily ramp check 1 followed by ramp checks 2, 3 and 4 at increasing numbers of hours flown (ramp check 4 is at 625 hours). Service checks 1, 2 and 3 start at 1,060 hours going up to 3,875 hours with an Intercheck at 6,250 hours and Major check every 24,000 hours or five years, whichever comes first. These critical checks keep the aircraft safe and reliable but are complex and time-consuming. An aircraft out of service, even on a planned check, is not earning its keep and delays can be very costly, not least if hangar space is needed for the next aircraft due in for service. Even what may seem a small modification, in this image a cargo door warning light change is being fitted, needs extensive stripping out of panelling and other wiring and needs very careful planning.

Above left and below left: By the early 1990s and with an ever growing fleet, British Airways Engineering had to have more 747 maintenance capacity. Moving away from Heathrow became the preferred choice and land was secured at Cardiff airport to build a massive three hangar facility capable of carrying out all inter and major checks and modification/design changes such as Section 41 and other life extension works. Named British Airways Maintenance Cardiff (BAMC), the facility opened in June 1993 and was set up as a wholly-owned subsidiary of British Airways, although it reported into British Airways Engineering, itself now a US$1billion business centre with significant third party maintenance revenue earning capabilities. In future years BAMC would become a major part of that revenue generating capability given its then unparalleled engineering facilities and the most efficient production methods in the industry.

Above and opposite: BAMC would quickly be followed by another two new subsidiary companies, British Airways Avionics Engineering (BAAE) and British Airways Interior Engineering (BAIE). BAAE would be another state-of-the-art facility with the latest specialised test equipment and workshop facilities for complex avionic maintenance that would double the number of units then being produced at Heathrow. Costing only £22m, it was a very cost-effective investment and provided significant capacity for third-party work.

Left: By contrast in terms of cost, BAIE was to bring together under one roof specialist interior soft furnishings engineers to build, repair and manage the hundreds of aircraft interiors and thousands of aircraft seats that are flown across the world by British Airways each day of the year and take a considerable battering from constant use. Needing only large workshops and storage space rather than complex test rigs, the cost of operating such a facility was considerably more cost-effective to run in South Wales, with good motorway links to Heathrow, Gatwick and the Midlands, than near London. Combined with a highly skilled local labour force, BAIE's engineers excelled in the skill of making a refitted aircraft look like new. Even quite small fittings that had traditionally been thrown away and replaced with new, would be repaired for a fraction of the cost and be indistinguishable, sometimes better, than new. The South Wales engineering labour force that British Airways recruited and trained to work at BAMC, BAAE and BAIE more than met the company's decision and expectations in devolving those businesses.

Opposite: Within two years of opening BAMC was carrying out its first Section 41 extended life modification on a sixteen year old British Airways 747-200. The check was one of thirty six modifications carried out and ran parallel to a full major check so the aircraft was completely stripped out. The nose section needed particular care to replace the aircraft's 'skeleton', the individual frames comprising the main structure, without removing the fuselage skin. Inside the cabin areas, all fixtures and fittings were removed including the carbon fibre floor panels, being replaced with plywood while the works were carried out to remove the risk of damage from any loose rivets that might get stuck to the soles of the engineers' shoes. BAMC developed the 'superkit' philosophy i.e gathering everything together that was needed for a specific job before the aircraft even arrived. The Section 41 'superkit' comprised 12,000 individual parts – from entire fuselage frames to metal shims (but not rivets or fasteners) – and simplified and speeded up the work considerably with no waiting around for missing or out-of-stock parts. 'Superkits' are now widely applied across British Airways Engineering.

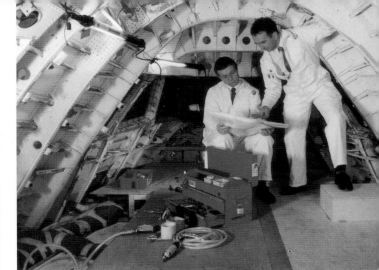

Opposite left: Life modifications were common across most of the legacy airlines in the 1990s as they sought to extend the useful lives of their respective aircraft. British Airways' 737s would undergo a major check plus that would include wing-root modifications, a complete strip out of all fittings down to the 'skeleton' of the aircraft and even removal of the tail-fin for corrosion control checks

Above and opposite right: To many people the Boeing 747 - 400 looks just like any other 747 variant except for its easily distinguishable winglets at the end of the -400 wings. They are not to increase lateral stability but to reduce drag and one way of doing that is to increase an aircraft's wing span. Less drag means less fuel used and designers are continuing to experiment to find the best aerodynamic effect. The 747 - 400 winglets are quite modest in size compared to the overall size of the aircraft although they are still six feet high, while later versions of the much smaller Airbus A320 have winglets so large they are called 'sharklets'.

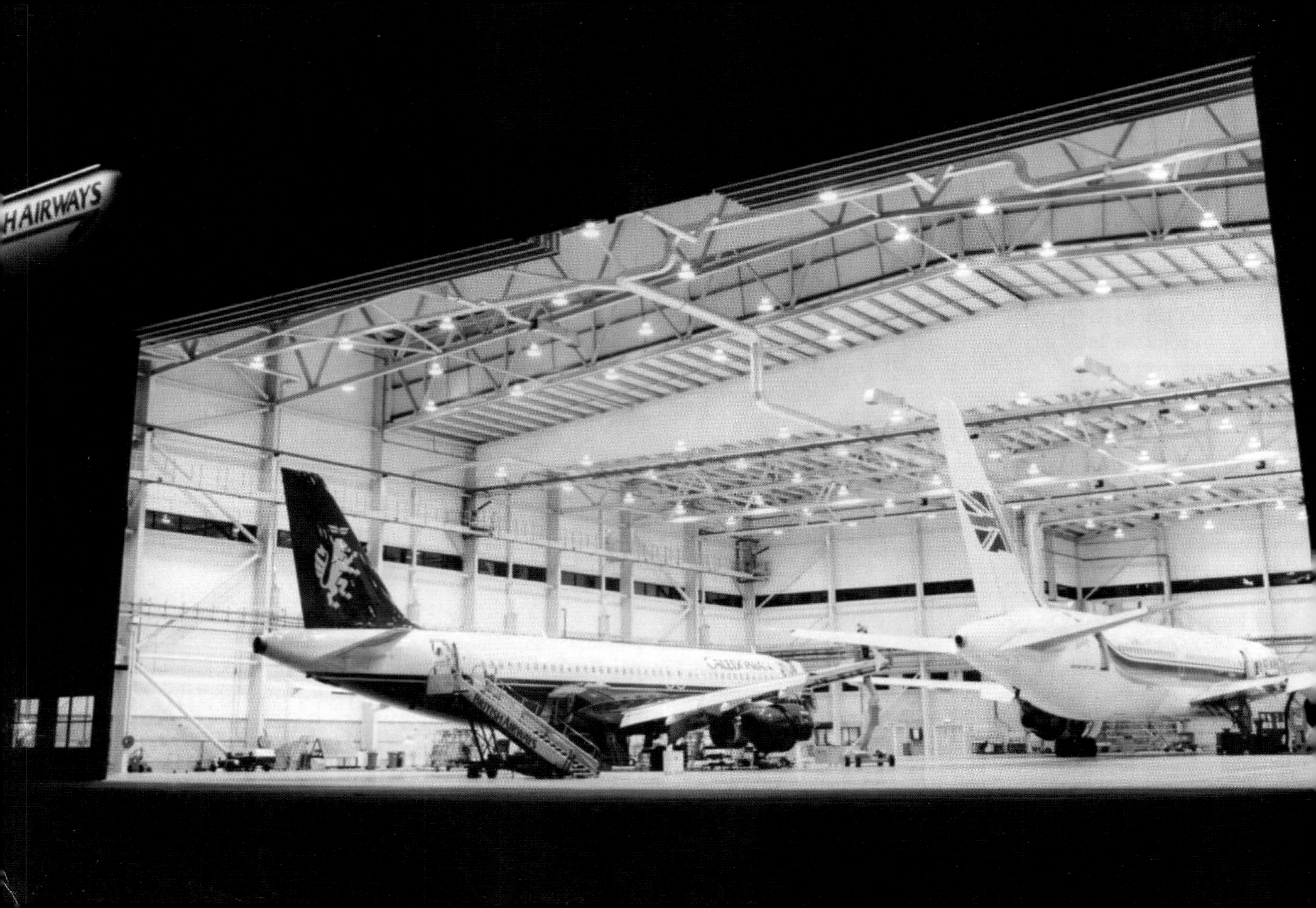
H AIRWAYS

This spread: Manchester has always been an important operating base for British Airways. By the mid 1990s the airport was developing fast with many new air services being introduced, particularly long-haul services by foreign airlines whose aircraft needed servicing. To improve its own fleet engineering capabilities but also to tap into this potentially lucrative third-party market, British Airways invested in a new 'super' hangar, super in that it was large enough to take even a 747-400 or the new Boeing 777-200 or a combination of up to five smaller aircraft types such as the Boeing 737 or Airbus A320. This was a largely revenue generating business decision that proved successful until the massive forced changes to the airline industry following the 9/11 terrorist attacks . With third-party business falling rapidly away, the hangar was closed and sold in 2002 and the engineering facilities scaled back.

Next spread: Just into its 20th anniversary British Airways took the decision to extend Concorde's life with an expectation of it continuing in service until at least 2007 and, possibly, even beyond 2015. This was an ambitious plan involving maintaining the integrity of the aircraft's structure and control systems as well as ensuring the retention of dedicated maintenance facilities, sufficient spare parts and engineering expertise; Concorde maintenance relied heavily on the skills of its engineers and had no push-button diagnostic testing facilities like many modern aircraft. Succession planning was a key issue to solve as older engineers retired and moved on. Spares were also potentially a problem as often none existed and had to be made from new, a very expensive exercise contributing to Concorde's annual maintenance cost of around £50m. Co-operation between British Airways and Air France as the only operators of the aircraft as well as the manufacturers, Aerospatiale and British Aerospace, were essential factors toward achieving the life-extension plan.

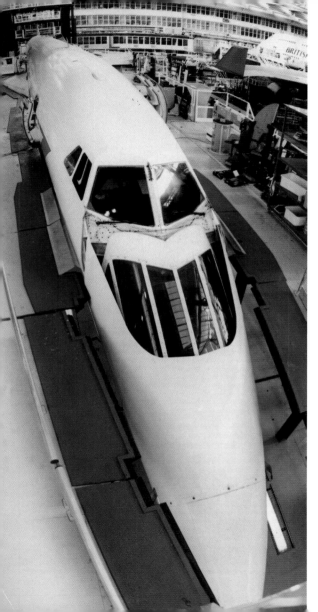

Above: While Concorde's engineers often had the skills to produce small spare parts, they relied on the manufacturers to build major components. In late 1993 Concorde G-BOAA became the first supersonic aircraft in the fleet to have a completely new rudder fitted. The original manufacturing jigs had been destroyed many years previously so a completely new production line had to be built. Nine new rudders were eventually built to replace those on the fleet of British Airways' five operational aircraft plus spares. The new rudders were made to a far higher specification than the originals bearing in mind the anticipated life-extension programme for the aircraft that would be initiated just three years later.

This page and opposite right: Boeing's new 777 - 200 aircraft was delivered to British Airways in November 1995. Described as the quietest, largest and most rigorously tested twin-jet aircraft in aviation history, it was as eagerly awaited as the ground-breaking Concorde and the 747 had been in their day. The aircraft's development had been the result of close co-operation between Boeing, the engine manufacturer General Electric, and British Airways Engineering and was expected to deliver the lowest seat mile costs for an aircraft of its size and with a noise footprint a third smaller than a 747. Some called it an aircraft built by engineers for engineers and its structure, systems and engines were very advanced. The aircraft's diagnostic systems were so good they allowed almost instant problem solving and its reliability was exceptional, to the point where it would regularly fly to Australia and back with no technical defects en-route whatsoever.

Left, opposite: The GE90 engine was a big, powerful engine by any standards delivering over fifty per cent more power than the Rolls Royce RB211-524H engines on British Airways' 747-400s. When tested by General Electric using a 747 as a flying test-bed, the engine cowling alone dwarfed the 747's standard engines.

Above: Moving the GE90 was a challenge given the engine's sheer size. Measuring some thirteen feet in diameter, almost the same width as the fuselage of a 737-400 aircraft, it was a big load to transport.

Opposite: Advances in electronic engineering had revolutionised aircraft flight decks in the 1980s and early 1990s, but the launch of the 777-200 heralded a new era in the evolution of flight navigation and control equipment. Flat panel Active Matrix Liquid Crystal Displays (AMLCD) did away with 20th century illuminated cathode-ray tube equipment to provide high information displays grouped by shade and colour with the ability to highlight critical information. Common in today's tablet computers and smart phones, AMLCD displays have since become standard equipment on 21st century civil airliners and a central component in avionics engineering maintenance.

Left: 1997 saw British Airways launch its 'World Images' livery, or coat of many colours as some would have it. With a multiplicity of multi-cultural art from across the world, it represented a major challenge to the engineers who had to paint the artworks on to each aircraft's tail fin. Some designs were more complex than others with this image designed by the artists Emmly and Martha Masanabo from the N'debele tribe of Southern Africa, being one of the easier ones to replicate. Using twelve different colours totalling 180 litres of paint and hand rollers, it would take four days to paint the 2000 square feet of a 747's tail section.

Next spread: Towards the end of the century British Airways' fleet began to take on the shape that took it well into the 21st century. The original 747-100s would all be sold as would thirty-four of its 757s to be converted into cargo freighters. The 777-200 fleet would be joined by the extended range version but with Rolls Royce Trent 895 engines and fifty nine Airbus A320s were ordered as well as two Airbus A318s and over thirty A319s. The DC10s were sold. Apart from some smaller aircraft types such as the BAE ATPs operating regional services, British Airways' fleet were consolidating around Boeing and Airbus aircraft types.

This spread: Concorde, of course, had a special place in the British Airways' fleet and was planned to fly on well into the 21st century. An announcement that it was to be grounded from August 2000 following the tragic, unexplained loss of an Air France Concorde at Paris, began an intensive programme of investigation into solutions to allow the aircraft to fly again. Led by British Airways Engineering, the solution was to line Concorde's fuel tanks with flexible Kevlar to increase the tanks' puncture resistance. This was a slow, difficult and claustrophobic task requiring engineers to crawl into the narrow fuel tanks openings to fit the linings into very restrictive spaces. This and other safety-related modifications would cost £17m plus another £14m to upgrade the cabin interiors. It is a great credit to Concorde's engineers that the aircraft was able to return to services between London and New York in November 2001 and Barbados a month later. By the summer of 2003, however, the serious decline in demand for a premium service following the 9/11 attack forced the withdrawal of Concorde services on its scheduled routes and charter services from 24 October 2003.

Next spread left: Customising its aircraft interiors had been a British Airways' practice for many decades, but it came at a high price when ageing aircraft were sold on and needed first to be re-configured back to industry standards. This 747 First cabin from the early 2000s was a very luxurious product for its time but a liability when trying to sell an aircraft.

Right: British Airways Engineering's maintenance base at Glasgow (BAMG) had undergone many changes over the years carrying out heavy maintenance checks on the airline's short-haul aircraft, principally the 737 in later years. With the growth of the Airbus fleet, BAMG gained its own airworthiness accreditation (JAR 145) in order to be able to take on Airbus heavy maintenance work. This was a critical step in securing BAMG's future role as a world class provider of shorthaul heavy maintenance for British Airways. (Jim Davies)

Next spread left: British Airways Engineering at Gatwick (BAEG) had undergone its own massive change following the strategic decision to change it from a 'hub' operation to 'point-to-point' services. At its peak in the 1990s, Gatwick occupied five hangars and two ramp satellites operating thirty four wide-body aircraft and nearly sixty narrow-bodied with major maintenance lines for the DC10, 737 and 757 and several major third-party contracts. By 2006, only Hangar 6 was left although this was given a major refurbishment a year later as part of a modernisation programme to improve the working environment and concentrate facilities on five work stations for light maintenance and aircraft 'casualty' work

Above left: By 2006/7 the burgeoning growth in new airline operators was creating an incentive for new third-party airline maintenance and repair facilities (MRO) to be established. This was almost a reverse of the situation a decade earlier with demand now driving supply as well as the competition from new Asian markets such as China encouraging its own MRO start-ups. Even ATC Lasham in the UK was competing for business in facilities that look more like a flying club hangar than British Airways Maintenance Cardiff's state-of-the-art facilities. State-of-the-art facilities, however, often incur increased costs and it could be a delicate balance to find the right blend of cost, expertise and price to bring in the business.

Above right: Gatwick Engineering was in a strong position to attract third-party work as it occupied one of the two maintenance hangars (Hangar 6) able to take on both planned input and ad hoc work. British Airways' own subsidiary, Open Skies, positioned all its aircraft to Gatwick from its Paris base for their service checks and any cabin modifications. Hard times had encouraged the Hangar 6 team to be flexible and proactive and by 2011 their third-party targets were regularly being broken with a range of airlines from Aer Lingus to Flybe being taken on.

Right: The casualty unit at Heathrow, in BEA's old 'Cathedral' hangar, was also searching for third-party work. This Qantas Airbus A380 does not quite fit but as the aircraft only needed an engine change then the tail could stay outside.

Above: The merger of British Airways and Iberia was expected to create more value under IAG's ownership than the two companies could deliver individually. British Airways Engineering and Iberia Maintenance were tasked to generate an extra EUR65m over IAG's first five years with a focus on controlling costs and growing revenue at a minimal additional investment while improving quality and safety by sharing best practice. An early win, together with the parts inventory house, AJW Aviation, was to maintain a component-care package for Thomas Cook Airlines' Airbus fleet that neither of the two airlines could have delivered independently.

Left and opposite: The introduction of the Airbus A380 raised its own particular problems due to the huge size of the aircraft. While some airline operators of the aircraft had to build new maintenance hangars, TBA at Heathrow was sufficiently flexible to allow the twenty four tonne hangar door lintels to be raised by four metres and a simple notch cut out to accommodate the aircraft's tail. Airbus sent one of their prototype A380 aircraft in February 2012 to try out the modified hangar.

A bit like buses, both the Boeing 787 and Airbus A380 arrived at British Airways in the same week (June 27 and 4 July 2013 respectively), shown together in this image at the British Airways Engineering base at Heathrow.

TBA's West Pen neatly fits British Airways first A380, G-XLEA.

This spread: British Airways Maintenance Cardiff's hangars also had some modifications works to fit the 787 although these images were taken in May 2016 as aircraft G-ZBJA came in for its first planned maintenance 'B' check. G-ZBJA is the first of a total of forty two 787s on order for British Airways. (Gareth Richards)

Overleaf: Both the A380 and 787 carried out a series of proving flights before beginning their commercial operations. This image is of British Airways first A380, G-XLEA, at Heathrow testing the Terminal 5 ground equipment procedures before a flight to Frankfurt on 13th August 2013.